Cambridge, 23 September 2020

WARFA D

Warfare was a recurrent phenomenon of fundamental importance throughout Roman history. Its scale and form varied across time and place, but it had wide-ranging impacts on politics, society and economy. This book focuses on important themes in the interplay between warfare and these broader contexts, including attitudes to war and peace, the values associated with military service, the role of material resources, military mutiny and civil war, and social and cultural aspects of the military. It also examines experiences of warfare, focusing on approaches to Roman battle and the impact of war on civilians. Importantly and distinctively, these different themes are traced across a millennium of Roman history from the Republic through to the end of Late Antiquity in the early seventh century, with a view to highlighting important continuities and changes across Roman history, and alerting readers to valuable but often less familiar material from the empire's final centuries.

DOUG LEE is a Professor of Ancient History in the Department of Classics and Archaeology at the University of Nottingham. Previous books include *Information and Frontiers: Roman Foreign Relations in Late Antiquity* (Cambridge, 1993) and *War in Late Antiquity: A Social History* (2007).

KEY THEMES IN ANCIENT HISTORY

Key Themes in Ancient History aims to provide readable, informed and original studies of various basic topics, designed in the first instance for students and teachers of Classics and Ancient History, but also for those engaged in related disciplines. Each volume is devoted to a general theme in Greek, Roman, or where appropriate, Graeco-Roman history, or to some salient aspect or aspects of it. Besides indicating the state of current research in the relevant area, authors seek to show how the theme is significant for our own as well as ancient culture and society. It is hoped that these original, thematic volumes will encourage and stimulate promising new developments in teaching and research in ancient history.

Other books in the series

Death-Ritual and Social Structure in Classical Antiquity, by Ian Morris
978 0 521 37465 1 (hardback) 978 0 521 37611 2 (paperback)
Literacy and Orality in Ancient Greece, by Rosalind Thomas
978 0 521 37346 3 (hardback) 978 0 521 37742 3 (paperback)
Slavery and Society at Rome, by Keith Bradley
978 0 521 37287 9 (hardback) 978 0 521 37887 1 (paperback)
Law, Violence, and Community in Classical Athens, by David Cohen
978 0 521 38167 3 (hardback) 978 0 521 38837 5 (paperback)
Public Order in Ancient Rome, by Wilfried Nippel
978 0 521 38327 1 (hardback) 978 0 521 38749 1 (paperback)
Friendship in the Classical World, by David Konstan
978 0 521 45402 5 (hardback) 978 0 521 45998 3 (paperback)
Sport and Society in Ancient Greece, by Mark Golden
978 0 521 49698 8 (hardback) 978 0 521 49790 9 (paperback)
Food and Society in Classical Antiquity, by Peter Garnsey
978 0 521 64182 1 (hardback) 978 0 521 64588 1 (paperback)
Banking and Business in the Roman World, by Jean Andreau
978 0 521 38031 7 (hardback) 978 0 521 38932 7 (paperback)

Roman Law in Context, by David Johnston
978 0 521 63046 7 (hardback) 978 0 521 63961 3 (paperback)
Religions of the Ancient Greeks, by Simon Price
978 0 521 38201 4 (hardback) 978 0 521 38867 2 (paperback)
Christianity and Roman Society, by Gillian Clark
978 0 521 63310 9 (hardback) 978 0 521 63386 4 (paperback)
Trade in Classical Antiquity, by Neville Morley
978 0 521 63279 9 (hardback) 978 0 521 63416 8 (paperback)
Technology and Culture in Greek and Roman Antiquity, by Serafina Cuomo
978 0 521 81073 9 (hardback) 978 0 521 00903 4 (paperback)
Law and Crime in the Roman World, by Jill Harries
978 0 521 82820 8 (hardback) 978 0 521 53532 8 (paperback)
The Social History of Roman Art, by Peter Stewart
978 0 521 81632 8 (hardback) 978 0 52101659 9 (paperback)
Ancient Greek Political Thought in Practice, by Paul Cartledge
978 0 521 45455 1 (hardback) 978 0 521 45595 4 (paperback)
Asceticism in the Graeco-Roman World, by Richard Finn OP
978 0 521 86281 3 (hardback) 978 0 521 68154 4 (paperback)
Domestic Space and Social Organisation in Classical Antiquity, by Lisa C. Nevett
978 0 521 78336 1 (hardback) 978 0 521 78945 5 (paperback)
Money in Classical Antiquity, by Sitta von Reden
978 0 521 45337 0 (hardback) 978 0 521 45952 5 (paperback)
Geography in Classical Antiquity, by Daniela Dueck and Kai Brodersen
978 0 521 19788 5 (hardback) 978 0 521 12025 8 (paperback)
Space and Society in the Greek and Roman Worlds, by Michael Scott
978 1 107 00915 8 (hardback) 978 1 107 40150 1 (paperback)
Studying Gender in Classical Antiquity, by Lin Foxhall
978 0 521 55318 6 (hardback) 978 0 521 55739 9 (paperback)
The Ancient Jews from Alexander to Muhammad, by Seth Schwartz
978 1 107 04127 1 (hardback) 978 1 107 66929 1 (paperback)
Language and Society in the Greek and Roman Worlds, by James Clackson
978 0 521 19235 4 (hardback) 978 0 521 14066 9 (paperback)
The Ancient City, by Arjan Zuiderhoek
978 0 521 19835 6 (hardback) 978 0 521 16601 0 (paperback) 979
Science Writing in Greco-Roman Antiquity, by Liba Taub
978 0 521 11370 0 (hardback) 978 0 521 13063 9 (paperback)
Politics in the Roman Republic, by Henrik Mouritsen
978 1 07 03188 3 (hardback) 978 1 107 65133 3 (paperback)
Roman Political Thought, by Jed W. Atkins
978 11 07 10700 7 (hardback) 9781107514553 (paperback)
Empire and Political Cultures in the Roman World, by Emma Dench
978 0 521 81072 2 (hardback) 978 0 521 00901 0 (paperback)

WARFARE IN THE ROMAN WORLD

A. D. LEE
University of Nottingham

CAMBRIDGE
UNIVERSITY PRESS

CAMBRIDGE
UNIVERSITY PRESS

University Printing House, Cambridge CB2 8BS, United Kingdom

One Liberty Plaza, 20th Floor, New York, NY 10006, USA

477 Williamstown Road, Port Melbourne, VIC 3207, Australia

314–321, 3rd Floor, Plot 3, Splendor Forum, Jasola District Centre, New Delhi – 110025, India

79 Anson Road, #06–04/06, Singapore 079906

Cambridge University Press is part of the University of Cambridge.

It furthers the University's mission by disseminating knowledge in the pursuit of education, learning, and research at the highest international levels of excellence.

www.cambridge.org
Information on this title: www.cambridge.org/9781107014282
DOI: 10.1017/9781139013680

First published 2020

A catalogue record for this publication is available from the British Library.

Library of Congress Cataloging-in-Publication Data
NAMES: Lee, A. D., author.
TITLE: Warfare in the Roman World / A.D. Lee, University of Nottingham.
DESCRIPTION: Cambridge ; New York, NY : Cambridge University Press, 2020. |
SERIES: Key themes in ancient history | Includes bibliographical references and index.
IDENTIFIERS: LCCN 2020022966 (print) | LCCN 2020022967 (ebook) |
ISBN 9781107014282 (hardback) | ISBN 9781107638280 (paperback) |
ISBN 9781139013680 (epub)
SUBJECTS: LCSH: Military art and science–Rome–History. | Rome–History, Military. |
Rome–Social life and customs.
CLASSIFICATION: LCC U35 .L443 2020 (print) | LCC U35 (ebook) |
DDC 355.020937–dc23
LC record available at https://lccn.loc.gov/2020022966
LC ebook record available at https://lccn.loc.gov/2020022967

ISBN 978-1-107-01428-2 Hardback
ISBN 978-1-107-63828-0 Paperback

For my family

Contents

Maps and Illustrations

Maps

Illustrations

Abbreviations

References to ancient sources generally follow abbreviations in the *Oxford Classical Dictionary* or (for late Roman sources) the *Prosopography of the Later Roman Empire*.

Other abbreviations (e.g., journals) follow those in the *Oxford Classical Dictionary* or the *Oxford Dictionary of Late Antiquity*.

Preface

There is no shortage of books on Roman warfare and the Roman army, so the production of yet another warrants explanation. This volume differs from many existing publications in this field in a number of respects. First, it is primarily a social history of warfare in the Roman world – in other words, it is less concerned with the operational details of wars and battles (although these do receive some attention, especially in the final chapter), and more interested in the relationship between military affairs and other areas of Roman life, whether that be politics, the economy, society or religion. This broader focus reflects the series remit, but also arises from a concern to contribute to the ongoing integration of Roman military history into Roman historical studies more broadly. Because of this focus, the approach is a thematic one, although the Introduction includes a narrative overview of major Roman wars and an outline of the organisational evolution of Roman military forces, for purposes of orientation. Secondly, the chronological range of this volume is broader than that of many previous studies, which have tended to focus on the Republic and Principate, with the occasional nod towards developments in the early years of Late Antiquity. This volume aims to give due attention to all three of these main periods, with Late Antiquity defined as the period from the mid-third century to the early seventh century. This is partly due to personal interest – my own primary area of expertise is Late Antiquity – but it is also the case that inclusion of a late antique perspective on the themes considered in this volume has the potential to offer novel insights into familiar issues, particularly given the rich but often less well-known material available from that period.

Pursuing this subject across a millennium of history – from the mid-fourth century BC to the early seventh century AD – is ambitious, particularly within the constraints of this particular series, whose volumes are rather shorter than a typical monograph. This has forced me to be selective in the themes considered and in the depth of treatment, but

I hope to have struck an appropriate balance, while also providing sufficient guidance to further reading. I have tried to take account of relevant recent literature, but given the time periods and range of subjects covered, I am bound to have overlooked some items. I particularly regret not having been able to take account of Simon James' new book on Dura-Europos.

My thanks to Peter Garnsey, Paul Cartledge and Michael Sharp for inviting me to contribute to this series, for their exemplary patience in waiting for me to deliver and for their comments on the text. Much of the work for this book was undertaken during an extended period of leave granted by the University of Nottingham in 2013–14, with support from the Dean's Fund, and then a further semester's leave in 2016–17, for which I thank the University, as also the School of Humanities for help with the costs of indexing. John Rich kindly gave very detailed feedback on earlier drafts of the first four chapters, correcting various misconceptions about aspects of Republican and Augustan history, and offering valuable suggestions for improvement, for all of which I am greatly indebted to him. More recently, Philip Rance and Greg Woolf read a complete draft and provided very helpful comment on a range matters, for which I am most grateful to both. For help with specific queries, my thanks to Kate Gilliver, Emily Kneebone, Simon Malloch, Conor Whately and George Woudhuysen; I am grateful to Wayne Boucher, Hal Churchman, Daria Lanzuolo, Berna Polat and Thomas Zühmer for assistance with images, as also to Philippa Jevons for her production of the index, to Barbara Wilson for the copy-editing, and to Sarah Starkey and Ishwarya Mathavan for overseeing production of the book. My thinking on a number of subjects that feature in this book has been stimulated by questions and comments from MA students at Nottingham in ancient warfare modules in which I have been involved over a number of years. Recent PhD students have also benefited my work on aspects of this book through our discussions of their research, so it is a pleasure to acknowledge Stuart McCunn, Nikki Rollason, Robert Stone, Rebecca Usherwood and Michael Wuk.

The members of my family have taken a patient and encouraging interest in the completion of this book. In the period since my previous book, the family has expanded significantly as our three children have married and in some cases added a new generation, enriching our life together in so many ways. So my thanks and love go to James, Sarah and Madison, Phil, Becs and Sebastian, and Naomi and Tom – and, as always, to my wife Anna, for her love, forbearance and common sense. This book is dedicated to them all.

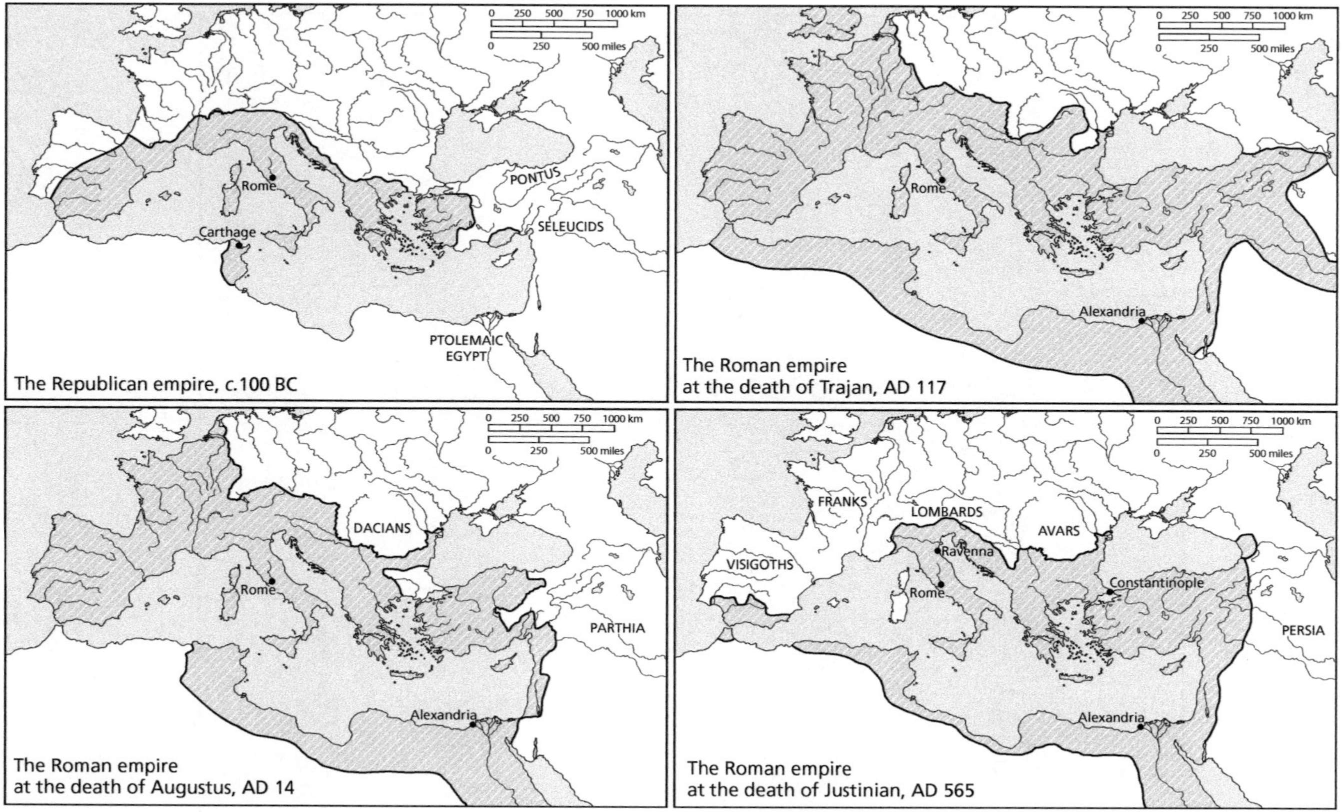

Map 1 The expansion and contraction of the Roman empire

Map 2 Roman Italy

Map 3 The Roman west

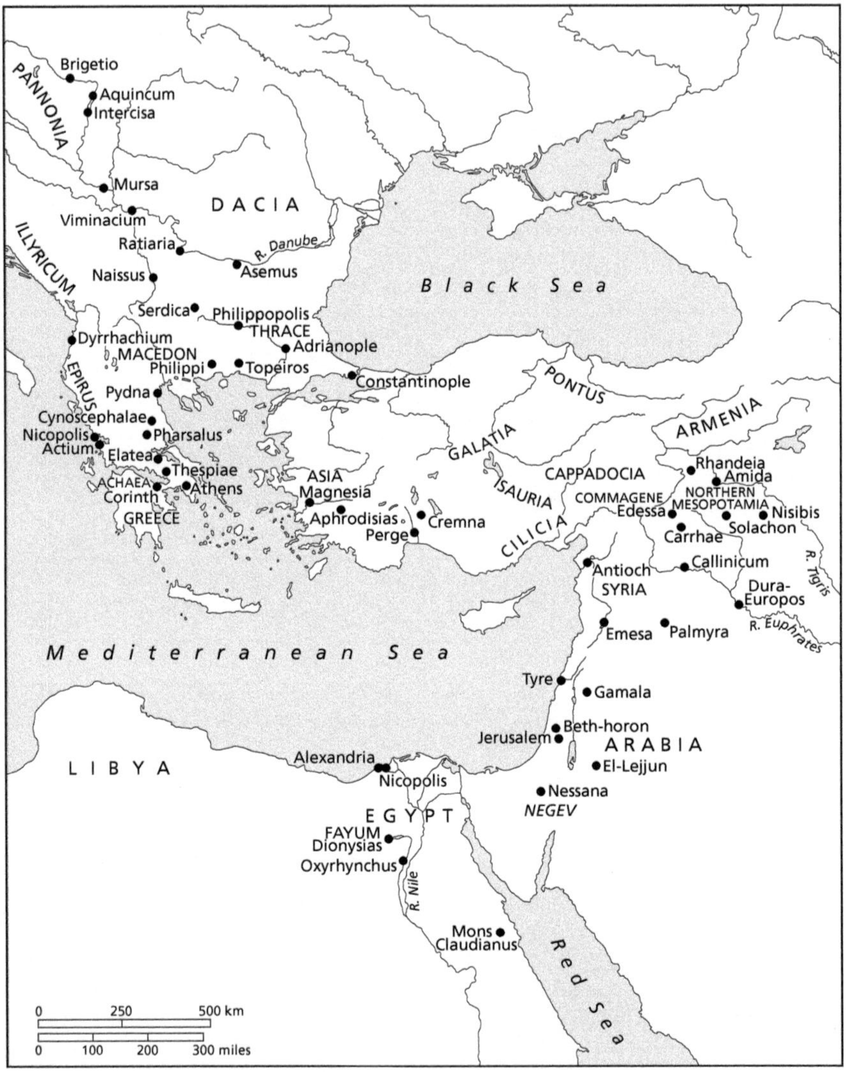

Map 4 The Roman east

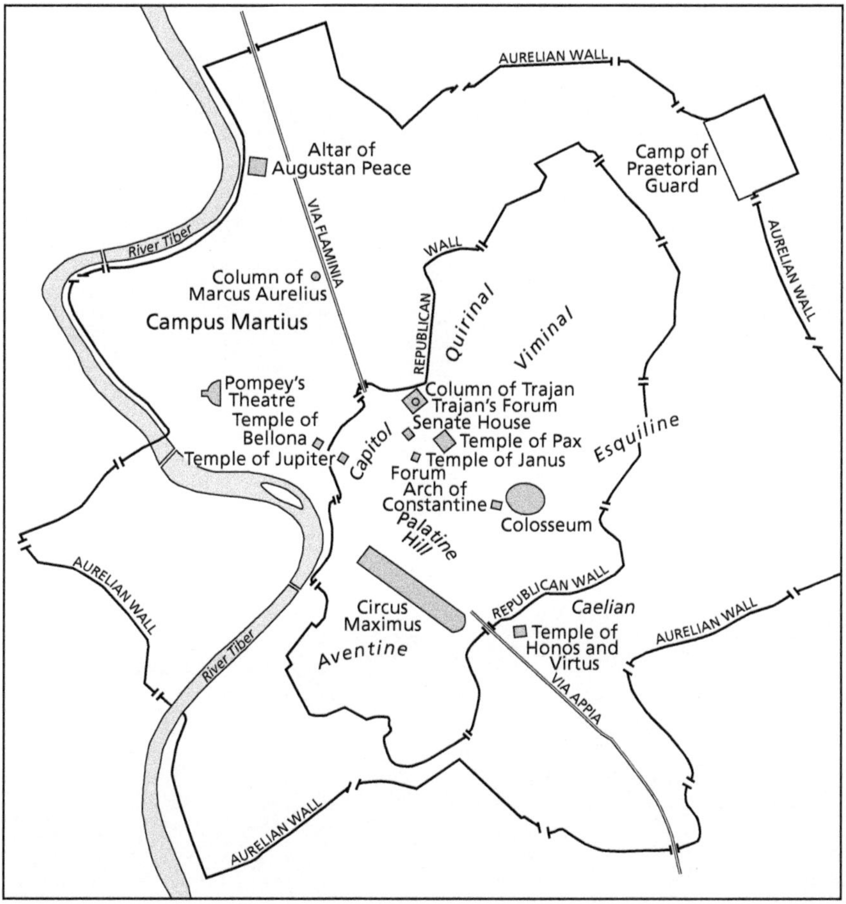

Map 5 The city of Rome

Introduction

In the mid-160s, the satirist Lucian produced a short treatise on how to write history.[1] In the introduction, he explains that he was prompted to address this subject by the way in which the recent Roman campaign against Parthia had spawned a spate of historical writing: 'Everybody has become a Thucydides or a Herodotus or a Xenophon, and apparently the old saying "War is the father of all things" [Heracleitus *fr.* 53] has been proved true, judging by the number of historians it has produced at one go' (*Hist. conscr.* 2). Although Lucian proceeds to deliver an amusing but devastating critique of these efforts, his work serves as a reminder that writing about war was deeply embedded in Roman culture, partly reflecting the enduring influence of Herodotus and Thucydides on classical historiography, but even more so the familiarity of military conflict in the Roman world and its close relationship with power. All this makes warfare a key theme of fundamental importance in Roman history. It also means that it is a very large subject, requiring difficult choices as to approach and coverage, especially for a volume in this series. The approach adopted here is thematic, and for that reason this Introduction aims to provide contextual orientation on a number of fronts. After explaining the book's parameters, a concise narrative overview of major wars in Roman history is provided, followed by an outline of the organisational evolution of Roman military forces and finally an introduction to the most important ancient sources and evidence for warfare in the Roman world.

1 Rules of Engagement

'Warfare in the Roman world' may seem like a self-explanatory title, so it is important to unpack the implications of the key terms as they are understood in this volume. First, 'warfare' obviously includes different forms of

[1] For discussion of the work, see Jones 1986: 59–67.

ancient military conflict – battles, sieges, raiding and the like – but it also extends to consideration of the institutions that made war possible and the agents of conflict – armies, officers and soldiers. Second, this volume treats 'the Roman world' as more than a geographical expression: it is taken here as encompassing its social, political and economic life. A major criticism of military studies as a field within Roman history has been its 'virtual "ghettoization"'[2] – that is, its marginalisation from broader scholarly debates and developments – and one of the underlying aims of this book is to continue efforts to integrate Roman military history into mainstream study of the Roman world by placing Roman warfare in its wider social, political and economic context.[3] Above all, this means thinking about the impact of warfare on a number of fronts beyond the immediately military. Finally, the term 'Roman' demands particular comment – a term less easy to define than one might perhaps think. Its chronological dimension is the most straightforward. In the context of antiquity, it can be defined as the period from the emergence of a settlement at Rome probably sometime in the eighth century BC until Late Antiquity, usually regarded as ending with the disruption of the Islamic invasions in the early seventh century.[4] Its geographical dimension changed significantly over time as Roman power gradually expanded (primarily through warfare, of course) to encompass the Italian peninsula, then the western Mediterranean, then the whole Mediterranean and much of its hinterland, before contracting over the course of Late Antiquity (Map 1).

The expansion of Roman territorial power over time, in turn, had fundamental implications for the meaning of Roman identity, which proved to be a very flexible identifier. As increasing numbers of inhabitants of the Italian peninsula and then the wider Mediterranean were incorporated into Rome's territorial empire, the term 'Roman' came to refer not just to the inhabitants of the city of Rome but, formally, to those living elsewhere who gained the privilege of Roman citizenship, and informally, increasingly to all inhabitants of the empire – a situation eventually confirmed by the extension of Roman citizenship to virtually all free inhabitants in the early third century. Matters were further complicated during Late Antiquity as individuals of barbarian origin found employment in the Roman military, in some cases gaining prominent positions of

[2] James 2002: 2.

[3] Ideally, this would include extended consideration of major neighbouring states and peoples, but constraints of space have limited the scope for this in this volume.

[4] All dates are AD unless otherwise indicated or unclear from context.

command. It is therefore important to bear in mind that the term 'Roman', when used with reference to people, meant different things at different times and in different places – and also that a person self-identifying as 'Roman' might at the same time also self-identify as a Samnite, Gaul, Athenian or Frank.[5] Nor, equally importantly, was the ability to speak Latin always a requirement for being a 'Roman'. While it would have been expected during the earlier centuries of Roman history, and being able to communicate in Latin strengthened such a claim and brought many practical advantages, the great majority of inhabitants of the Roman empire at its fullest extent had a different first language (e.g., Greek, Celtic, Punic, to name only the most widely spoken) and may only have ever acquired a smattering of Latin – though significantly, the Roman armed forces were an important context in which knowledge of Latin was promoted.[6]

Having defined the chronological range of Roman history as starting with the emergence of a settlement at Rome probably in the eighth century, it is important to add that, as implied by the word 'probably', reliable historical sources for the early centuries of Roman history are very limited – and so this book takes the mid-fourth century BC as its chronological starting parameter, both because sources are better and because this was when Roman territorial expansion began in earnest. This means that its coverage involves three broad periods of Roman history – the Republic (or more strictly the middle and late Republic), the Principate, and Late Antiquity. As is so often the case, this periodisation is defined primarily with reference to politics. The political character of the Republic has been the subject of intense debate in recent scholarship, focusing on where the balance lay between the influence of democratic elements and the senatorial elite, but whatever one's views on that question, the Republic's institutions were intended to guard against one-man rule, and it therefore stands in clear contrast to the autocratic regime which the first emperor Augustus established in the final decades of the first century BC. However, mindful of the anti-monarchical traditions of the Republic and the fate of his adoptive father Julius Caesar, Augustus referred to himself as *princeps* – leading citizen – and so one common

[5] Consider, e.g., the late Roman epitaph from Aquincum on the Danube which begins with the claim *Francus ego civis Romanus miles in armis* ('I [am] a Frankish citizen [and] a Roman soldier in arms') (*ILS* 2814).

[6] For helpful discussions of Roman identity, see (among others) Brennan 1998b: 191–3, Woolf 2012: ch. 14, Mattingly 2014, James 2014, Dench 2018; for Late Antiquity, Greatrex 2000, Mathisen 2006, Conant 2012: 1–9.

designation for the regime which he established and which enjoyed stability until the early third century is the Principate.[7] Late Antiquity refers to the period from the mid-third century, when the Roman state was in serious danger of fragmenting, to the early seventh century, when the Islamic invasions reshaped the Roman state in significant ways – with the loss of the western half of the empire during the fifth century a further fundamental development during that period. The recovery of the empire from the third-century crisis owed much to the emergence of a new type of emperor. In contrast to Augustus and his successors during the Principate, who were almost all drawn from the senatorial elite, these were men from military backgrounds (and therefore lower in social status) who undertook major reorganisation of significant aspects of the Roman state, most obviously an expanded military establishment and bureaucracy.

Although these broad divisions of Roman history are traditionally demarcated with reference to major political changes, it is relevant to the focus of this book that they also map onto significant changes in military organisation. More detail about the evolution of Roman military forces will be provided below, but at the risk of oversimplifying developments, the headline features are as follows. The core of the Republic's armed forces was a citizen militia in which individuals were liable to sixteen years of service overall, but could normally expect discharge after six years continuous service, whereas Augustus established a standing army, with individuals typically serving continuously for a minimum of twenty years (from AD 5). This fundamental change to the basis of military service continued into Late Antiquity, but – unsurprisingly, given their military backgrounds – emperors in the early fourth century introduced major organisational changes to the armed forces which meant that important features of the military in Late Antiquity looked very different from the military of the Principate.

The thematic approach adopted here starts with two chapters whose focus is the relationship between warfare and ideas. Chapter 1 begins by considering Roman attitudes to war and peace, before turning to the related subject of the Roman ideology of victory and responses to its counterpoint of defeat, while Chapter 2 examines the Roman ideology of military service and its changing relationship to citizenship and property ownership, and then discusses the Roman ideal of courage. In Chapter 3,

[7] This term is used here in preference to 'Early Empire' because of the ambiguity of the word 'empire', which can refer to both a territorial entity, such as the Republic acquired, and a political system in which supreme power resides with an emperor.

the focus shifts to the more practical issues of manpower and money – recruitment and the demographic impact of mobilisation, and how the Roman state financed war-making and the material benefits which it brought the state. In Chapter 4 issues relating to the theme of authority and allegiances are discussed, with consideration of generalship in theory and practice, the incidence and causes of military mutiny and the dilemmas of civil war. Chapter 5 focuses on the military as an institution in the context of Roman society and questions of identity, examining soldiers as a community, their relations with wider society, and the religious dimension of military life. Chapter 6's theme of communication and culture is pursued in a number of directions – the role of warfare in cultural interchange at the level of weaponry and tactics, the cultural impact of the military's presence in the Roman world, and the question of literacy in the armed forces and its implications. The final chapter focuses on the experience of warfare from a number of different perspectives – that of soldiers in the context of set-piece battles, with particular reference to debate about the 'face of battle', and that of non-combatants in a range of contexts, above all siege warfare, but also raiding and protracted wars. In considering these different subjects, the discussion aims to give balanced consideration to developments in the Republic, Principate and Late Antiquity, with a view to highlighting significant continuities and changes in the impact of warfare across the trajectory of Roman history.

2 Warfare in Roman History: A Strategic Overview

Warfare was a significant feature throughout the history of the Roman world and the purpose of this section is to provide an overview of its incidence across the relevant centuries, especially for orientation of readers who may be less familiar with the detail of Roman history.[8]

The Republic

The Republic was broadly a period of territorial expansion through war, though the rate of expansion was by no means uniform. In the early Republic, Rome controlled only its immediate hinterland in central Italy, as it contested dominance of the region with other Latin communities. By

[8] For narrative overviews with a military focus (to the late fifth century), see Mackay 2007 and Roth 2009, and for Late Antiquity to the seventh century, Elton 2018.

the mid-fourth century BC Rome had established its pre-eminence in western central Italy, and next confronted Samnites and associated Italic groups who controlled the regions further south in the peninsula. Through a series of protracted wars against these very determined opponents, Rome established itself as the dominant state across central Italy, on the basis of which it began to exert its influence further south and northwards, gradually extending its network of subordinate allies. The Greek colony of Tarentum in the south resisted the spread of Roman influence by forming an alliance with Pyrrhus, the ruler of Epirus in the western Balkans, who brought his forces across to Italy in 280 BC. He achieved some successes against the Romans, but was eventually forced to withdraw from the peninsula, so that by the 260s Rome controlled the whole of the Italian peninsula south of the Po valley, thereby making it a major power in the western Mediterranean.[9]

It is perhaps unsurprising that this soon led to confrontation with the other major power in the western Mediterranean, Carthage. Originally a Phoenician settlement on the northern coast of Africa, Carthage had by the early third century developed its influence in north Africa, Spain and Sicily. In 264 Rome intervened in Sicily and initiated the first of three wars with Carthage, which involved a heavy investment of human and material resources and which marked another major step in the expansion of Roman power (the so-called Punic Wars, after the Latin name *Poeni* for the Phoenician Carthaginians). The first war with Carthage was fought predominantly at sea, a medium where one would have expected Carthage, as a seaborne power, to have the upper hand. However, Rome proved adept at developing its naval capabilities, and although the war dragged on for more than two decades, with both sides suffering setbacks, Rome eventually inflicted a decisive naval defeat on Carthage off Sicily in 241, which forced Carthage to agree to stiff peace terms. Carthaginian resentment fuelled a desire to reassert its influence, which found an outlet in Spain until the able general Hannibal led a land invasion of Italy in 218. A string of early victories, culminating in the crushing Roman defeat at Cannae in 216, looked like it might achieve Hannibal's aim of undermining the loyalties of Rome's Italian allies and forcing Rome to negotiate, but although some allies defected, the majority did not, and Rome's superior manpower resources eventually won the day, with an even more

[9] Cornell 1995: chs. 12, 14.

stringent peace imposed on Carthage in 201. The third and final war in the early 140s was a one-sided coda.[10]

By 200 BC, then, the Roman state had sustained a century and a half of almost continuous war, mostly against very resolute enemies in the form of the Samnites and the Carthaginians, and the final defeat of Hannibal left Rome as the dominant power in the western Mediterranean. During the second century BC, the pattern of warfare fluctuated. On the one hand, there were some regions that required regular lower-intensity campaigning to establish or re-assert Roman control, notably in Spain and in northern Italy where Celtic and Ligurian tribes had long been a serious threat.[11] On the other hand, there were occasional wars, each lasting just a few years, above all against the Hellenistic kingdoms of Macedon and the Seleucids in the eastern Mediterranean – successor states which emerged from the empire of Alexander the Great at the end of the fourth century BC. Although the rulers of these states controlled significant resources and had well-organised military forces, they nonetheless proved unable to match the Romans when it came to war. Rome embarked on a campaign against Macedon as early as 200 BC because Macedon's ruler, Philip V, had previously allied himself with Hannibal, and the Romans now sought revenge. Roman forces defeated Philip in the Balkans in the early 190s, above all at the battle of Cynoscephalae (197), before taking on the Seleucid ruler Antiochus III in Greece and Anatolia, where Roman military power was again demonstrated in a number of decisive battles, notably at Magnesia (190). Further conflict with Macedon in the late 170s resulted in a decisive Roman victory at Pydna in 168 and the end of the Macedonian kingdom. Although Rome did not immediately take territorial control of Macedon and Greece, and although Roman control of Seleucid Syria and Ptolemaic Egypt was not imposed until the mid to late first century BC, there was no doubting that by the mid-second century it had become the dominant power in the eastern Mediterranean, whose wishes local rulers and states sought to follow.[12] The late second century saw Rome waging war in north Africa and then confronting fresh challenges from Celtic tribes in southern Gaul, as well as Germanic tribes, with Roman forces experiencing some major defeats before eventually prevailing.[13]

[10] Hoyos 2011, Rosenstein 2012a: 53–70, 119–75, 233–7. For Cannae, see Daly 2002.

[11] Spain: Richardson 1986; Italy: Rosenstein 2012a: 176–9.

[12] Rosenstein 2012a: chs. 5–6.

[13] Steel 2013: 26–31.

Although there were further periodic bouts of warfare during the first century BC which extended Roman territory and influence, above all through the eastern campaigns of Pompey in the 60s and the Gallic campaigns of Julius Caesar in the 50s, the first century BC was one in which civil war also loomed large for the first time in Roman history, ensuring that apart from the 90s there was significant conflict in every decade to the end of the Republic. The first episode of civil war involved the revolt of Rome's Italian allies in 91 BC, aggrieved that they contributed so much to Roman military success yet received, in their view, too little reward – above all denial of the privileges of Roman citizenship. Rome eventually resolved this conflict through a combination of military action and concessions, but not before the so-called Social War (the war against Rome's allies or *socii*)[14] spilled over into conflict between two of Rome's leading generals, Marius and Sulla, and their supporters, during the 80s BC. Sulla was victorious, although there was further civil war in Spain when the Roman general Sertorius revolted against the Sullan regime. A further, even more wide-ranging round of civil war between politically ambitious generals was initiated in 49 BC when Julius Caesar invaded Italy with the legions from his Gallic campaigns, taking on the forces loyal to the senate and commanded by Pompey. The ensuing conflict, from which Caesar emerged victorious in 45 BC, ranged across the Mediterranean world, with campaigns in Spain, north Africa, and the Balkans.[15] Caesar's murder in 44 BC triggered a new round of civil war, initially between Caesar's supporters and his assassins (resolved in favour of the former at the battle of Philippi in 42 BC – although with further aftershocks arising from Sextus Pompey's control of Sicily until 36 BC) and then between Caesar's supporters themselves, above all his lieutenant Mark Antony and his young heir Octavian. The latter was eventually triumphant at the battle of Actium in 31 BC, four years later adopting the name Augustus and establishing himself as the unchallenged ruler of the Roman world.[16]

The Principate

Given the almost constant warfare in which Augustus had been engaged during the first decade and a half of his adult life (44–30 BC), it would have been understandable if he had opted for a more relaxed existence once he had secured supreme power in the Roman world. However, Augustus and his generals are credited with adding more territory to the Roman state

[14] Dart 2014. [15] Steel 2013: chs. 4, 6. [16] Osgood 2006, Richardson 2012: chs. 2–3.

than anyone before or after, including Egypt, northwest Spain, the Alps and much of the Balkans – and it seems that he also had ambitions to expand beyond the Rhine into Germany. This military activity has been explained most persuasively with reference to the demands of internal politics. Following the defeat of Antony, it was essential to Augustus' political position for him to retain control of the armed forces, and the simplest initial justification for doing so was the pacification of provinces which were insecure either because of the risk of internal rebellion or from external threat. In 27 BC Augustus committed to this task for a ten-year period of authority, which was then repeatedly extended at ten-year intervals until his death in AD 14 – a process which has been deftly characterised as 'making the emergency permanent'.[17] This commitment to pacification meant regular warfare, especially in the first half of Augustus' reign, but any plans for establishing permanent control in Germany were halted in AD 9 when three legions operating beyond the Rhine were ambushed and massacred in the Teutoburg Forest – the infamous Varian disaster, so-called after their ill-fated commander Varus[18] – and this no doubt encouraged Augustus' immediate successors to adopt a more cautious military approach.

What is striking about the Principate, however, is that the period as a whole post-Augustus saw very little further imperial expansion. The two major exceptions were the decision of the emperor Claudius to initiate a campaign to conquer Britain in AD 43 and the emperor Trajan's conquest of Dacia, north of the lower Danube, in the early years of the second century. Given the lack of a strategic or economic rationale for adding Britain, Claudius' decision is best seen as a case of a decidedly unmilitary emperor seeking to strengthen his legitimacy with an easy military success. As an experienced general, Trajan had no such need, and a punitive war rather than conquest of Dacia may have been his original intention.[19]

Limited territorial expansion, however, does not mean that were no other significant instances of warfare during the Principate. After the conquest of Dacia, Trajan embarked on a less successful attempt to conquer Parthia, to the east. The Parthians, an originally nomadic people who had taken over the eastern territories of the Seleucid kingdom during the second century BC, emerged as a major neighbour of the Roman state when the latter extended its reach into the Levant in the first century BC, and had demonstrated their military capabilities in defeating the Roman general Crassus at Carrhae in northern Mesopotamia in 53 BC. Julius

[17] Rich 2012a. [18] Wells 2003. [19] Levick 1990: ch. 13, Bennett 2000: 89.

Caesar was planning a major expedition against them at the time of his death, but (as in other matters) Augustus resisted the temptation to follow in his adoptive father's footsteps; instead he contented himself with achieving the diplomatic success of persuading the Parthians to return the legionary standards captured at Carrhae, presenting this event in 20 BC as virtually equivalent to a great military victory.[20] A century or more later, however, Trajan invaded Parthia in 115 and reached the Persian Gulf before insurgencies forced him to withdraw. Further campaigns against Parthia were undertaken under the emperors Marcus Aurelius in the 160s and Septimius Severus in the 190s, with some territorial gains in northern Mesopotamia by the latter.[21] However, these episodes of increasing Roman military superiority served to destabilise the Parthian Arsacid regime, which was overthrown in the 220s by the Sasanian family who established a new dynasty which was to prove a formidable opponent of the Roman state during Late Antiquity. Germanic and Sarmatian groups also became growing threats on the Danube frontier in the later second century, with Marcus Aurelius having to spend significant time campaigning against them in the 170s.[22]

As in the first century BC, however, external wars were not the only manifestation of warfare in the Roman world. There were also instances of civil war, the first occurring in 68–69, the second in 192–3. The former – the so-called 'Year of the Four Emperors' – was precipitated by the overthrow of the childless Nero, the last member of the Julio-Claudian dynasty established by Augustus, by one provincial governor with armed forces at his disposal, which then prompted others in similar positions to make bids for power. The eventual winner was Vespasian, who restored internal stability. Likewise, the overthrow of the childless Commodus in 192 prompted competition for power by leading senators with provincial armies under their command, from which Septimus Severus emerged as victor.[23] Internal conflict during the Principate also took the form of provincial rebellions. The most serious of these were those associated with the Jews, with three major instances: the war in Judaea in 66–73, the Jewish revolt in Egypt, Libya and Mesopotamia in 115, and the Bar Kokhba revolt in Judaea under Hadrian in the early 130s. All of these required the commitment of substantial military forces to achieve their suppression. There were also significant instances of rebellion in the

[20] Cornwell 2017: 125–38. [21] Bennett 2000: ch. 13, Birley 1999: ch. 13, 2000a: chs. 6–7.
[22] Birley 2000a: ch. 8. [23] Birley 1999: chs. 10–11, Levick 2017: ch. 4.

western provinces, notably the revolt in Britain by Boudicca in the early 60s and the revolt of the Gallic noble Civilis in 68.[24]

Late Antiquity

As already noted, the third century saw the emergence of a major new challenge to the Roman state from the east in the form of Sasanian Persia. The new dynasty speedily marshalled its resources to challenge the Roman empire, launching destructive invasions into Syria in 252 and 260. The latter was notable as the first occasion in which a Roman emperor (Valerian) was captured by an enemy. At the same time, the empire faced more serious threats along the Rhine and the Danube, where Goths, Alamanni and other groups made major incursions into Roman territory, achieving another first when the emperor Decius was killed in battle against the Goths on the lower Danube in 251. It was these vicissitudes of the mid-third century which accelerated the gradual replacement of emperors of aristocratic senatorial background by men of somewhat lower social status with military backgrounds and experience. These 'soldier emperors' spent their reigns on the frontiers rather than in the city of Rome and gradually re-established the empire's stability, driving back northern invaders and reasserting Roman authority in the east against Sasanian Persia. A major defeat was inflicted on the Sasanian king in 298, allowing Diocletian to impose a chastening peace on them in 299, which included the addition of some new territory east of the Tigris and effectively fixed the frontier until 363. However, the third century was also notable for an efflorescence of civil war as different armies championed their commander against others and significant regions of the empire pursued 'separatist' courses. Among other instances, the emperor Aurelian had to suppress the breakaway state headed by the eastern city of Palmyra which seized control of many eastern provinces in the 260s, before turning to do the same to the breakaway 'Gallic empire' in the west.[25]

Although Diocletian tried to circumvent the impulse to civil war by sharing power with others in the arrangement known as the Tetrarchy ('rule by four'), his experiment ultimately failed in the face of the ambitions of Constantine, who engaged in successful civil war between 312 and 324, first in the west and then in the east. His victories established stability again, allowing the foundation of a new capital in the east at

[24] Jewish revolts: Goodman 2008: chs. 12–13; western revolts: Dyson 1971.
[25] Potter 2014: chs. 6–7.

Constantinople, which increasingly became the focus of imperial attention, with the role of the city of Rome in the empire's affairs ever more peripheral. However, that stability did not last, as his three sons faced challenges from one another and from other contenders for the throne, especially in the 350s (with that of the military officer Magnentius the most serious). Internal dissension also encouraged neighbouring groups to seize the opportunity to invade or raid Roman territory, notably in Gaul where Alamanni and Franks took advantage of the disruption of Magnentius' usurpation. Constantine's nephew Julian is credited with the restoration of order in Gaul during this period, notably through his victory over an Alamannic confederation in a pitched battle at Strasbourg in north-east Gaul (357). During the decades after Constantine's death, Sasanian Persia also presented a renewed challenge to the empire under the energetic leadership of king Shapur II. Constantius II met this challenge astutely with an approach based on strengthened frontier defences and avoidance of pitched battles, but his cousin Julian squandered the resulting stability when he invaded Persia in 363 and was killed in the process. The resulting peace imposed on his successor served to alleviate the humiliation of 299 as far as the Persians were concerned, and helped ensure stability in the east for the next century and a half.[26]

However, the stabilisation of the eastern frontier was soon offset by destabilisation of the Danube frontier, where, in the 370s, the emperor Valens faced a renewed problem with the Goths. Valens had already weathered an early challenge to his position from the would-be usurper Procopius, but in 376 he was presented with the dilemma of whether he should agree to the request of large numbers of Goths seeking admission to the empire to escape the inroads of nomadic Huns from further north. Valens decided to admit one group, who then rebelled against Roman authority and began raiding the provinces south of the lower Danube. Eventually in 378 Valens confronted them near Adrianople where, against expectations, the Goths won a crushing victory which one contemporary likened to Hannibal's at Cannae in 216 BC (Amm. Marc. 31.13.19). Valens himself died in the battle, and his successor Theodosius proved unable to expel the Goths from Roman territory, finally agreeing in 382 to their being settled on imperial land, in return for providing soldiers – troops referred to in the sources as *foederati* (lit. 'allies', with the implication of a degree of independence).[27] Meanwhile in the west, there was a recurrence of civil war through the rebellion of military commanders, first

[26] Potter 2014: chs. 9–10, 12–13.

[27] Lenski 2002a: chs. 2–3, 6, Kulikowski 2007: chs. 5–6.

Magnus Maximus in 383 and then Arbogast in 392. Twice Theodosius marched west to suppress the rebellions, only to die unexpectedly in 395, leaving his two young sons to succeed him. They faced a further revolt of Gothic soldiers, led by their rebellious general Alaric, which culminated in 410 in the Gothic sack of the city of Rome, although this was more an act of frustration than of strategy. The early years of the fifth century also witnessed invasions across the Rhine and Danube by other barbarian groups, of whom the most serious proved to be the Vandals.[28]

The Vandals became a serious threat because, after making their way south through Gaul and Spain during the 420s, they occupied north Africa, the wealthiest region in the western half of the empire. Despite repeated imperial attempts to dislodge them during the mid-fifth century, they remained in control of the region and its tax revenues, thereby contributing significantly to the erosion of imperial power in the west.[29] Other barbarian groups, especially Goths and Franks, gradually gained control of other regions of the west, until the final few western emperors exercised authority over only the Italian peninsula, with the last of them deposed in 476. The eastern half of the empire had meanwhile been containing major challenges from the Huns on the lower Danube, and then after the death of their most famous leader Attila in 453 and the breakup of his empire, threats from some of the barbarian groups previously under his authority, notably other groups of Goths. Imperial power in the east was further weakened by the unpopularity of the emperor Zeno, who faced internal military challenges on a number of fronts during the 470s and 480s. His successor Anastasius had to deal with rebellion in the Anatolian region of Isauria during the 490s, then renewed war with Persia in the first decade of the sixth century, and then a further internal rebellion led by the military officer Vitalian in the middle years of the 510s.[30] It was only under his successor but one, Justinian, that imperial military power experienced something of a revival, albeit short-lived. Despite an inconclusive war with Persia early in his reign, Justinian oversaw the reconquest of north Africa from the Vandals in 533–4 and a concerted attempt to dislodge the Goths from Italy. Although imperial forces had seemed to be on the verge of victory in 540, however, the latter war dragged on until the early 550s, not least because a fresh round of war with Persia from

[28] Kulikowski 2007: chs. 7–8, Lee 2013a: 26–7, 110–14.

[29] Heather 2005: ch. 6, Merrills & Miles 2010: chs. 2–3, 5.

[30] Lee 2013a: chs. 6, 8.

540 forced Justinian to divert military resources to the east. There was also a military rebellion in north Africa to deal with.[31]

Justinian's successor, Justin II, renewed war with Persia in 572, which proved a major error of judgement since it went disastrously at first and dragged on for the next two decades. On the lower Danube, the nomadic Avars became a new and serious threat to imperial security. With the conclusion of peace with Persia in 591, the emperor Maurice was able to focus efforts on dealing with the Avars and Slavic raiders during the 590s, only for a revolt by troops serving on the lower Danube in 602 to result in Maurice's overthrow and replacement by one of the officers, Phocas.[32] This in turn provided Persia with the opportunity to renew warfare against the empire, this time achieving a degree of success which had previously eluded it. By 626 Persian forces controlled most of the empire's eastern provinces, including the economically vital Egypt, and had reached the Bosporus where they threatened to capture Constantinople itself – which would almost certainly have signalled the end of the empire. Persian successes had been aided by civil war within the Roman empire, as the governor of Africa, Heraclius, sailed east to challenge Phocas in 610. That challenge was successful, and over the next decade Heraclius began to rebuild Roman military capabilities to the point where in the mid-620s he launched a successful counter-offensive which achieved victory. However, Heraclius' triumph of 628 proved short-lived. The long war between the two powers had weakened both, so that when Arab forces under the banner of Islam emerged from the Arabian peninsula in the 630s, the Sasanian dynasty proved unable to resist and succumbed completely, while Heraclius suffered a major defeat in Palestine which resulted in the loss of many eastern provinces. With their territory reduced to a rump of Anatolia and a limited part of the eastern Balkans, and their economic resources severely reduced by the loss of Egypt, emperors were forced to rethink their whole approach to military organisation, so that the empire of the mid-seventh century and beyond – referred to in modern scholarship as the Byzantine Empire – was very different from that of Late Antiquity.[33]

3 The Evolution of Roman Military Forces

The expression 'the Roman army' can unhelpfully suggest an unchanging, monolithic entity,[34] so it is important to appreciate that Roman military

[31] Lee 2013a: chs. 12, 14, Heather 2018. [32] Whitby 1988: 24–7, 165–9.
[33] Kaegi 2003, Howard-Johnston 2010: chs. 14–15, Haldon 2016. [34] Cf. James 2002: 38–9.

forces comprised multiple elements which underwent significant organisational and associated changes at various stages over the centuries from the Republic through to Late Antiquity. This section outlines the most important of those changes, with a view to providing a framework within which to situate the themes discussed in subsequent chapters. At the same time, these changes themselves were also among the most significant impacts of warfare across Roman history.

Roman military forces during the Republic comprised two main categories of troops: citizen soldiers who served in legions, and troops provided by Italian allies, although in the final two centuries BC significant use was also made of non-Italian troops. Citizens were originally conscripted into the legions on an annual basis, with the number called up dependent on anticipated military commitments and individuals then returning home after the campaigning season. With those commitments increasingly involving service overseas during the middle Republic, however, some legions remained in service for a number of years. In principle individuals were liable for sixteen years of service overall, but they could normally expect discharge after six years continuous service.[35] Military service was not, then, generally a full-time permanent form of employment in this period of Roman history, although the manpower demands of the civil wars in the final decades of the Republic did result in some troops serving for more extended periods which anticipated Augustan reforms.

The organisation of troops serving in a legion underwent some important changes over the course of the Republic. During the middle Republican period, legionary soldiers were arranged into tactical units known as maniples, but by the late Republican period this had given way to organisation by cohorts. The size of maniples (lit. 'handfuls') depended on the category of soldier. Recruits were assigned to one of four categories depending on age and experience: the youngest were placed in the light-armed *velites* ('swift ones') who acted as skirmishing troops in front of the other three successive categories, comprising the *hastati* ('spear bearers'), the *principes* ('chief men') and the *triarii* ('third liners'). These were more heavily armed, with large shields, breastplate, helmet and greaves, sword and spear, the one difference being that the *triarii* had a thrusting spear (*hasta*) rather than a javelin (*pilum*) – a relic of the phalanx organisation from the early centuries of Roman history. Maniples consisted of either 120 men (*hastati* and *principes*) or 60 (*triarii*), and the great advantage that they offered over the more rigid phalanx that they replaced was their

[35] Brunt 1962: 80–81.

greater flexibility and manoeuvrability, as demonstrated in the Roman victory over the Macedonian phalanx at Cynoscephalae in 197 BC. Under the manipular system, each legion consisted of 30 maniples. By the first century BC, however, the standard tactical unit had become the cohort, a larger unit of 480 men, comprising 6 centuries (which consisted of 80 men, rather than the 100 that one might otherwise expect from the name). Under this arrangement, the legion comprised 10 cohorts, so the shift from maniples to cohorts seems to have been one which sought a better balance between flexibility and solidity – and experimentation with this appears to have been occurring from as early as the Hannibalic War. Alongside this organisational change, there was also greater uniformity of equipment, with the thrusting spear of the *triarii* being discarded in favour of the javelin.[36]

Troops contributed by the Italian allies were of central importance to Roman military success during the Republic, since by the late third century BC they constituted more than half of the forces at the disposal of the Roman state. According to Polybius (3.107.12), allied infantry in 216 were comparable in number to those in the legions, while allied cavalry was three times as numerous as Roman cavalry, although these proportions probably varied somewhat over time.[37] Italian allies contributed troops according to the individual agreements which different communities had concluded with the Roman state, as recorded in the document known as the *formula togatorum*.[38] Allied infantry units were known as *alae*, literally 'wings', because they were positioned on the flanks of legions, while allied cavalry units were known as *turmae*. It is generally thought that the allies were organised and equipped in a manner little different from the legions,[39] so when, after the Social War of the early first century BC and the concession of citizenship by Rome to the Italian allied communities, their units became legions, the process of adaptation will have been relatively straightforward.

Roman use of non-Italian troops during the final two centuries BC has received much less attention in modern treatments of Republican warfare, but even if their numbers are difficult to quantify, the evidence suggests that their use was extensive. The troops in question are not those provided

[36] Keppie 1984: 33–6, 63–7, Rosenstein 2012a: 94–8 for clear overviews; Oakley 1998: 451–76 for more detailed discussion of Polybius and Livy on maniples.

[37] Brunt 1987: 677–86, Rich 1983: 321–3.

[38] Lit. 'list of toga-wearers', where *togatus* probably indicates those who wore adult attire and were therefore of military age (Prag 2010: 19–20 for references).

[39] Keppie 1984: 22, Rosenstein 2012a: 98.

by non-Italian allies, but rather those levied from provincials (e.g., in Sicily and Asia). There was no fixed term for them, although *auxilia externa* ('foreign supporting forces') is used in some texts and is a convenient designation. Troops levied from provinces might be used on active campaigning and for garrison duty, while coastal communities might be expected to provide ships and their crews. Moreover, it has been argued that 'after the Social War when the Italian *socii* no longer constituted a distinct part of the Roman army, . . . the *auxilia* [*externa*] now constituted the principal secondary branch of the Roman army'.[40]

Longer overseas military campaigns and the incidence of civil war in the first century BC meant that some soldiers increasingly served more than the previously typical six years, but it was the emperor Augustus who formalised the shift from an army of part-time citizen conscripts into a full-time standing army.[41] At the end of the civil war in 30 BC there were as many as 60 legions under arms. Augustus more than halved their number to 28 legions and stationed them (together with auxiliary forces) in the provinces most in need of a military presence, with legionaries now serving (from AD 5) for a minimum of twenty years. 'In a radical departure from earlier practice, Augustus thereafter avoided raising new legions for specific occasions, opting instead to meet Rome's military needs from the permanent establishment, with units being redeployed as required.'[42] This change to a standing army is often referred to in modern studies as the creation of a 'professional' Roman army or its 'professionalisation', but while there are some aspects of this change which are perhaps consistent with that characterisation, the modernising assumptions implicit in language of this sort are unhelpful and such terminology is better avoided.[43] Nonetheless, there can be no doubt about the importance of the change, which had significant ramifications on a range of fronts, above all for political power, state finances and military identity.

Organisationally the legions continued to be based on the cohort principle. The legions were, however, now supplemented by units referred to as *auxilia*, which developed out of the *auxilia externa* of the later Republic. Like these and the Italian allies of the Republican period, they comprised troops who were not Roman citizens, but were drawn from the inhabitants of the provinces; like the Italian allies of the Republic, they came to constitute more than half of the empire's troops. *Auxilia externa* were heavily exploited during the civil wars of the late Republican period,

[40] Prag 2007: 76–87, 2010, 2011: 15–22, 2015 (quotation at 2010: 112). [41] Raaflaub 1980.
[42] Rich 2003: 345–6. [43] Lee 2020.

as competing generals sought to augment their forces at a time when citizen manpower was fully exploited, so that by the time of Augustus' victory auxiliary units had become an established part of the military landscape, particularly valued because they often provided troops with military skills which complemented the heavy infantry of the legions, whether it be more mobile forms of infantry, various types of cavalry or specialised missile troops.[44] Infantry units were organised in cohorts, and cavalry in *alae*, usually of about 500 men each (although larger units of up to 1,000 men appeared in the later first century AD). While the distinction between legions and *auxilia* began as one between citizen and non-citizen, this became progressively blurred in both directions over the second century.[45] Supplementing the *auxilia* were units usually referred to as *numeri*, recruited from tribal groups on the fringes of the empire, but not of any standard size or organisation.[46] The Principate also saw the establishment of a permanent military presence within the city of Rome itself, primarily in the form of the Praetorian Guard, initially intended to protect the person of the emperor in the city but gradually becoming an elite force which accompanied emperors on campaign. Initially numbering perhaps 4,500 men, the Guard grew in size to 10,000 by the end of the first century.[47]

The military upheavals of the third century resulted in major changes to the structure of the Roman military, albeit in gradual ways for the most part. In the latter stages of the Principate there had been a growing trend towards creating operational armies for specific campaigns by withdrawing detachments ('vexillations') from frontier legions, and this practice eventually led during the third century to the emergence of a more permanent force sometimes referred to as the *comitatus*, because it 'accompanied' the emperor on his increasingly constant campaigning. Although much of the unit terminology remained the same (e.g., legions, *auxilia*), the meaning of these terms often changed. In particular, military units in Late Antiquity tended to be much smaller than the legions of the Principate – typically, 1,000–1,500men – perhaps reflecting the break-up of multi-component legions into independent specialised units. Since the overall size of the armed forces increased, this meant a proliferation in the number of units, including the development of a greater number and variety of cavalry units during the mid-third and early fourth century. Perhaps the most significant change was a shift in the strategic distribution of the empire's armed

[44] Haynes 2013: 31–50. [45] Lavan 2019a: 31–3. [46] Southern 1989.
[47] Coulston 2018: 175–8.

forces. During the Principate, the overwhelming majority of units were based on the empire's peripheries. During the early fourth century, however, there was a major restructuring whereby the best units were drawn back from the frontiers to form strategically placed field armies (the so-called *comitatenses*, which evolved out of the *comitatus*), while those units which remained in frontier provinces (*limites*) came to be referred to as *limitanei*. By the late fourth century, there were five field armies in the east (two near Constantinople and one each in Thrace, Illyricum and the east), each under the command of a senior general known as a *magister militum* ('master of the soldiers'), with a similar pattern of regional armies in the west, although here the command was more centralised. In the sixth century, Justinian expanded their number by adding new field armies for Armenia, north Africa and Italy, reflecting his more ambitious plans for the empire. The creation of field armies placed a premium on mobility, so it is unsurprising that there was an increase in the number of cavalry units, especially mounted archers, reflecting the impact of the empire's encounters with Hunnic steppe nomads from the fifth century onwards (Figure 1). Nonetheless, there was no decisive shift towards mounted troops and infantry remained a core component of the armed forces in Late Antiquity. In addition to *comitatenses* and *limitanei*, another important category of unit from the late fourth century onwards was *foederati* (lit. 'allies'), usually drawn from barbarian groups settled within the empire in return for military service, while the sixth century saw the growing

Figure 1 Ivory relief showing sixth-century infantry and a mounted horse archer (Egypt, sixth century).
Rheinisches Landesmuseum Trier (T. Zühmer)

prominence of the retinues of generals referred to in the sources as *bucellarii* (lit. 'biscuit men', from soldiers' twice-baked bread rations, *bucellatum*). Finally, there was another category of ally drawn from groups residing beyond the limits of the empire (often designated in the sources as *symmachoi* or *enspondoi*), of which Arabs along the eastern frontier were perhaps the most significant.[48]

As this brief survey indicates, one of the hallmarks of Roman military forces through time was their adaptability to changing circumstances, as also a willingness on the part of the Roman state to exploit the manpower and military skills of others. These are important features in their own right which will receive further attention in later chapters, but they are also relevant to the broader issue of how Roman military forces are characterised. One popular metaphor is that of the Roman army as a smoothly running and invincible machine, but as a number of scholars have noted, this is an unhelpful analogy. It is unhelpful because it is based on modernising assumptions about the Roman army, it risks minimising the fact that Roman armed forces sometimes experienced defeat, it implicitly treats individual soldiers as automata, overlooking the many instances of cowardice, mutiny and rebellion – and its implication of rigidity and lack of initiative does not given sufficient weight to the adaptability of Roman military organisation over the *longue durée* and its pragmatic openness to drawing on the skills of others.[49]

4 Writing Roman Warfare

Because this book is concerned with warfare in its wider context, many different authors and genres of ancient literature contribute to its chapters – too many to survey here. However, as already noted, it is in Roman historical writing that warfare and its impact features particularly strongly and so this genre warrants special attention. The prominence of war in historical writing could be deduced even from what is known of works which have not survived, such as Coelius Antipater's history of the Hannibalic War (Cic. *Leg.* 1.6, *De or.* 2.54), Pliny the Elder's account of the Germanic wars of the early first century AD (Plin. *Ep.* 3.5.4), or John the Lydian's history of Justinian's early Persian wars (*De mag.* 3.28). The

[48] For armed forces during the transition from the Principate to Late Antiquity, see Whitby 2004, Garnsey 2019, and the important series of papers by Peter Brennan (1998a, 1998b, 1998c, 2007); for discussions of Roman military organisation which cover the full chronological range of Late Antiquity, see Elton 2007, Rance 2007; for *foederati*, see Stickler 2007; for Arab allies, Fisher 2011.

[49] Goldsworthy 1996: 283–6, James 2002: 9, 31–2, 2011a: 22–3.

historical works which are extant, whether in full or in part, are clearly significant sources, but it is important to exercise critical awareness when reading them and not simply take them at face value, while also taking account of other types of evidence. In what follows, a selection of the most important historical sources is introduced before turning to those other types of evidence.[50]

Before commenting on specific authors, however, there is a more general issue to be noted – the degree of reliability that can be attached to ancient historical accounts, whether of military affairs or other matters. While no modern scholar is likely to accept everything in an ancient account without caution or reservations about some details, some in recent decades have given particular weight to the literary dimensions of historical narratives arising from the influence of the rhetorical training that was such an important element of ancient education in the Greco-Roman world and have seen this as grounds for greater scepticism. Alertness to generic conventions and literary agenda is undoubtedly important when reading and evaluating ancient historical narratives, but it was also the case that ancient historians writing about near contemporary events faced the prospect of rebuttal and were aware of the need for their account to stand up to scrutiny. This is not to deny the limitations of the sources on which historians sometimes had to base their accounts, or the role of bias in their presentation of events, but neither should such constraints, nor the influence of stylistic considerations, be overplayed.[51]

Although historical writing in Latin began in the early second century BC, the first surviving substantial works are from the first century BC. The earliest important extant history is therefore that of a Greek historian, who took a particular interest in Rome's military affairs during the later third and first half of the second century BC. This was Polybius, whose promising political career in Achaea was one of the casualties of Roman imperial expansion. It would have been understandable if his forced relocation to Italy in 167 BC had led him to produce an embittered critique of Rome, but instead he wrote a work which sought to make sense of the transformation that the Greek world, as well as the wider Mediterranean, underwent at the hands of this previously modest Italian city state. This geopolitical revolution had a strong military dimension to

50 This discussion of ancient historical sources could easily have been expanded to include such writers as Sallust, Appian, Cassius Dio, Herodian and Agathias, all of whom feature significant episodes of Roman warfare. For a useful collation of bibliography on these authors as military historians, see Lendon 2017: 164–7.

51 See further Bosworth 2003, Damon 2007.

it, which meant that Polybius' history included a significant focus on war, but also an important discussion of the middle Republic's military institutions, as part of his famed analysis of the Roman state in Book 6. Although significant portions of Polybius' history have not survived, those that have provide important insights into major Roman military campaigns during the testing times of the Hannibalic War, as well as Roman confrontations with the Hellenistic kingdoms of the east. As a cavalry commander in Achaea before his exile to Rome, and subsequently as a member of the Roman senator Scipio Aemilianus' retinue during some of his campaigns, Polybius was able to write from personal familiarity with warfare (28.13.6, 38.19–22), while he is also known to have written a (lost) treatise on tactics (9.20.4).

The first Latin history in the classical tradition of which a significant portion has survived is that of Titus Livius (Livy). Writing during the reign of the first emperor Augustus, his ambition to provide a history of Rome from its foundation nearly eight centuries earlier meant that his work differed from that of Polybius in important ways besides language: it was much more ambitious in chronological scope, broader in subject matter and less concerned with answering a specific question in the way that Polybius was, and he is also not known to have had any military experience. Nonetheless, warfare features prominently and influenced the structure of the work,[52] and Livy also includes a description of the centuriate assembly, a political institution which had significance for Roman military organisation (1.43). Since his history stretched back many centuries before Livy's own lifetime, he was necessarily dependent on the works of earlier writers, including Polybius, although it was also the case that the earlier he went, the thinner such sources became, both in terms of reliability and existence. Like Polybius, significant portions of Livy have not survived: his first ten books cover the period to the early third century BC, while Books 21 to 45 run from the outbreak of the Hannibalic War to Rome's defeat of Macedonia in 168.[53]

A generation or so prior to Livy's own day had seen the production of one of the most important accounts of warfare during the Late Republic. These were Julius Caesar's narratives of his campaigns in Gaul during the 50s BC and of the civil war of 49–48 between Caesar, Pompey and the senate. His adoption of a plainer, unadorned manner of writing implied that he was not aspiring to write in the grand style of classical historiography and prompted contemporaries to refer to his works as *commentarii* –

[52] Levene 2009: 225. [53] See further Oakley 1997: 3–151, Levene 2010.

a term which could be translated as 'memoirs' or 'notebooks'. However, while their simpler language, and Caesar's use of the distancing third person to refer to himself, might tempt one to take them at face value, their simplicity conceals artful literary technique designed to further Caesar's political agenda. In the case of the *Gallic War*, the primary goal was making Rome aware of his military achievements, while in the *Civil War* it was a matter of justifying himself against his political enemies and identifying his cause with Rome's interests. Their value as military narratives from the perspective of an active participant in a position of command therefore needs to be balanced by awareness of these underlying concerns which may have had some influence on his presentation of military operations. At the same time, they do not appear to have resulted in significant distortions, and his accounts of both wars do not gloss over significant setbacks such as the failures at Gergovia and Dyrrhachium.[54]

The most important historical work written in the first century AD with respect to Roman warfare is the account of the Jewish revolt of 66–70 by the Jewish author Josephus – 'the Greek Livy' (Jer. *Ep.* 22.34) – both for its narratives of individual engagements and sieges, and also for its account of Roman military practices (*BJ* 3.70–109). Written in Greek, its opening paragraph mimics Thucydides' claim to have written about the greatest war of his lifetime, and Thucydides' influence can be seen elsewhere, both in episodic literary echoes and thematic elements. His experience as a participant in the early stages of the war on the Jewish side has the potential to enhance the value of his work, but this is offset by his subsequently siding with the Romans and needing to justify his change of allegiances, as well as his probable use of the *commentarii* of the emperors Vespasian and Titus, who also had their own agenda in the context of the civil war of 68–69.[55]

From the early second century AD there is the work of Tacitus. Although (tellingly) he lamented the fact that he had to record 'unimportant and trivial matters', in contrast to earlier Roman historians who wrote about 'great wars, the storming of cities, the defeat and capture of kings' (*Ann.* 4.32), and although his personal military experience was probably limited, he nevertheless provides important accounts of warfare

[54] See further Lendon 1999, Kraus 2007, 2009, Batstone and Damon 2006, Raaflaub 2009. From the same period Sallust's account of the war against Jugurtha in north Africa (112–105 BC) highlights the challenges Roman forces faced in dealing with guerrilla-style warfare, but this text has received much less scholarly attention than Caesar's from a military perspective. Paul 1984 provides a historical commentary.

[55] Rajak 2002, Roth 2016.

in the first century, above all in the surviving initial books of his *Histories*, covering the civil conflicts of the year 69, but also in portions of his *Annals* and biography of Agricola.[56]

The last great Roman historian writing in Latin was Ammianus Marcellinus, active in the second half of the fourth century. Like Caesar and Josephus, he wrote on the basis of personal experience of active military service, a point which he may have been aiming to emphasise in his famous concluding reference to himself as 'a former soldier' (31.16.9). Unlike Caesar and Josephus, Ammianus was not a commander, but as a staff officer to a senior general he had some experience of warfare during some of the most tumultuous decades of Roman history. In particular he saw active service on the empire's eastern frontier against Sasanian Persia, both in resisting the Persian invasion of 359 and participating in Julian's Persian expedition in 363. At the same time, the two great set-piece battles of the surviving half of his history – Julian's victory over the Alamanni at Strasbourg in 357 and Valens' defeat at the hands of the Goths at Adrianople in 378 – were events of which Ammianus was not an eyewitness, which perhaps helps explain their less satisfactory nature as accounts of battle. Moreover, Ammianus' unconcealed (though not unconditional) admiration for the emperor Julian adds a further element of difficulty to interpretation.[57]

The last great Greek historian writing in the classical tradition was Procopius, active in the mid-sixth century during the reign of the emperor Justinian, whose various military campaigns offered plenty of scope for classicising history, whether it be in the east against the Persians or in the west against the Vandals and the Goths. Not a military man himself (more likely a lawyer by training), Procopius nonetheless was an eyewitness of some of the events he describes by virtue of serving as secretary to Justinian's leading general, Belisarius. Like Josephus, the preface to Procopius' history of the *Wars* acknowledges Thucydides' influence (and Herodotus' too), while also intimating a competitive relationship with both. Although biases are evident in Procopius' history – above all, he amplified Belisarius' achievements in the early books and blamed setbacks on others, while being critical of many aspects of Justinian's reign – his

[56] For Tacitus' possible military posts, see Birley 2000b: 235, 237; as a military historian, Levene 2009.

[57] See further Matthews 1989: 279–303, Kelly 2008, Ross 2016.

substantial account survives in full and provides an extraordinary amount of circumstantial detail about warfare in his lifetime.[58]

A final narrative text which warrants specific mention is the early sixth-century *Chronicle* traditionally attributed to Joshua the Stylite.[59] The *Chronicle* provides a detailed account of the empire's war with Persia in the early years of the sixth century and is significant because, although not a soldier, the author was a contemporary living in the theatre of operations in northern Mesopotamia (probably a monk in Edessa) who also drew information about some events from eyewitnesses (25). Above all, however, he was writing outside the Graeco-Roman historical tradition. Although it has been suggested that he may have had some familiarity with Herodotus and Thucydides,[60] the author wrote in the Semitic language of Syriac (a dialect of Aramaic) and the late antique genre of the chronicle did not aspire to emulate Thucydides or his successors, so it is an account which is less encumbered by the traditions of classical historiography.

While ancient historical narratives of war are an essential resource for the study of warfare in the Roman world, many other genres of Roman literature, from letters to plays to poetry, include relevant incidental material that will feature in subsequent chapters but is too diverse and scattered for consideration here. However, there is one further category that warrants particular comment, namely military treatises. That of Vegetius is the best known, no doubt partly because of its enduring influence on medieval writing about warfare.[61] Although written in the late fourth or early fifth century and including much topical comment relevant to that period, this treatise was the work of a professed civilian who drew on earlier sources reflecting aspects of military practices from the Republic and Principate, and this can complicate its interpretation.[62] Earlier examples of this genre include Frontinus and Polyaenus on stratagems, Onasander on generalship and Arrian's military works,[63] while

[58] See further Greatrex 1998, Rance 2005, Whately 2016.

[59] Translation and commentary in Trombley & Watt 2000. It is now generally accepted that Joshua the Stylite was a later copyist, rather than the author, whose identity remains uncertain.

[60] Watt 1999. [61] Allmand 2011.

[62] Rance 2007: 344–5 (favouring a late fourth-century date; arguments for a fifth-century date in Charles 2007).

[63] See Campbell 2004 for a convenient compendium of extracts in translation, organised thematically, with discussion in Campbell 1987. For Polyaenus, see Wheeler 2010; for Onasander, Smith 1998, Chlup 2014; for Arrian's military works, Wheeler 1979, 2004.

from the end of the sixth century there is the rather different practical handbook for generals attributed to the emperor Maurice.[64]

All literary sources, whether historical or another genre, were bound to have elements of bias, whether conscious or not, which makes the contribution of documentary sources – inscriptions, papyri, ostraca and wooden writing tablets – all the more important.[65] Documentary sources are not without their own biases, but those biases are usually of a different sort, such as geographical and social variation. For warfare in the Roman world, inscriptions are the most significant category because they have the widest geographical distribution.[66] Epigraphic texts were inscribed on a range of media in antiquity, most commonly stone, pottery and metal. Inscriptions on metal were most vulnerable to recycling, because of their intrinsic value, and so far fewer have survived. However, there is an important subcategory of inscriptions on metal with special relevance to military affairs – the diplomas issued to soldiers serving in the *auxilia* and certain other branches of the armed forces on completion of their term of service (Figure 2). These were extremely valuable documents for the recipients because they proved the grant of Roman citizenship to them at the time of their discharge, and so were often buried for safe-keeping; more than 1,200 have been recovered in modern times. Their content provides important data regarding recruitment patterns and details of military units during the Principate.[67]

Stone-inscribed texts have survived in much larger quantities, with epitaphs the most common type – there are hundreds of thousands of them and unsurprisingly these include many instances of individuals who served in the armed forces, especially during the Principate. Some are very simple statements of name, age, military rank and/or unit, and dedicatee (s), while others offer considerable detail about life and career, sometimes with accompanying images. The latter were more likely to be those of individuals who had held some sort of rank. Such inscriptions have proved invaluable for reconstructing career patterns, as well as the movements of military units. A notable feature of soldiers' epitaphs is that, somewhat paradoxically, the great majority do not record the deaths of soldiers in battle. There are exceptions, but generally soldiers who died on the

64 Rance 2004a: 267–9, 2007: 347–8. The translation by Dennis (1984) will be superseded by Philip Rance's forthcoming translation and commentary.

65 Translations of many such documents relating to military affairs can be found in Fink 1971 (papyri), Campbell 1994 (inscriptions and papyri).

66 Overview of military inscriptions in Speidel 2014.

67 See further Cooley 2012a: 172–7, Speidel 2014: 337–40, Lavan 2019b.

Figure 2 Example of a bronze military diploma issued to auxiliary troops on completion of service and confirming the grant of citizenship and other privileges (Brigetio, 149) (13 × 21 cm; translation and discussion in Speidel 2014: 338).
Metropolitan Museum, New York/Wikimedia Commons

battlefield were more likely to have been placed in mass graves on site, with an appropriate cenotaph.[68] Another important category of inscription is soldiers' dedications to deities, which are a valuable source of information for religious practices in the military, as also imperial prouncements responding to civilian complaints about military requisitioning and the like.

Of other documentary media, papyri have survived in the greatest quantities. Papyrus was essentially an ancient form of paper, made from compressed layers of the pith of the papyrus reed, and was widely used

68 Hope 2003, Cooley 2012b.

throughout the Roman world in antiquity. However, unlike the durable materials on which inscriptions were written, papyrus was vulnerable to damp conditions, and so ancient papyrus documents have only survived in warm, dry locations, particularly in the Middle East, above all in parts of Egypt – but the quantities are nonetheless enormous, in the hundreds of thousands, with many still stored in museums and library repositories awaiting study. Because Egypt only became part of the Roman empire in the time of Augustus, any documentation is only relevant from that point on, and because Egypt was not a region of the empire usually exposed to serious military threat, much of the surviving material is of limited relevance to the subject of warfare. Nonetheless, because of Egypt's importance as a source of grain for Rome, a military presence was always maintained there, even if soldiers found themselves more involved in police work than fighting. From the fourth century there is a particularly valuable archive of letters, official and personal, kept by Abinnaeus, commander of a unit at the fort of Dionysias in the Fayum, which shed light on military–civilian interactions.[69] In some respects, however, it is another site, outside of Egypt, which has been more fruitful in papyri of military relevance – namely, Dura-Europos on the Middle Euphrates. Archaeologists found within the remains of this frontier fortress an invaluable cache of papyrus documents relating to a unit of Palmyrene archers stationed there in the early third century. The most discussed of these documents is the so-called *Feriale Duranum*, a calendar of official religious events celebrated by the military during the Principate.[70]

Other writing materials include ostraca (pieces of broken pottery), a cheap medium on which it was possible to write or scratch brief records. Many thousands have been found at the mining operation at Mons Claudianus in the Egyptian desert where there was a military presence, as also from the north Africa military outpost at Bu Njem, providing insights into practical aspects of military life and administration.[71] Wooden writing tablets were also widely used to record a variety of texts, both official and private, and some of the most interesting discoveries in recent decades have been caches of these from Vindolanda near Hadrian's Wall. These have shed light on such matters as logistics, literacy and the social life of troops stationed there in the second century AD.[72]

[69] For papyri in general, see Bagnall 2009; for papyri relating to the military in Egypt, see Haensch 2012, Palme 2007; for Abinnaeus, Bell *et al.* 1962, Barnes 1985.

[70] Welles *et al.* 1959.

[71] Marichal 1992, Adams 1994, Cuvigny 2003, Cooley 2012a: 275.

[72] Bowman 1994a.

Finally, there is the fundamental contribution of non-textual material evidence, encompassing archaeological sites (most obviously military camps, forts and other defensive structures), movable items (especially military equipment), and visual representations relating to warfare, such as reliefs on victory monuments.[73] In addition to illuminating changing approaches to defence, the layout of camps and forts enhances understanding of interaction among soldiers and between soldiers and civilians.[74] The study of military equipment aids insight into the ways Roman weaponry adapted to changing challenges and the mechanics of battle, among other things.[75] And visual representations can help to test – and sometimes disprove – statements made in textual sources, such as Vegetius' curious claim (*Mil.* 1.20) that soldiers ceased to wear armour in the late fourth century.[76] The fortress at Dura-Europos is unusually rich in all these different types of evidence: in addition to its papyri, it has proved invaluable for structures, military equipment, visual evidence and the remains of siege works.[77] For understandable reasons, the locations of and evidence from battlefields have proved less easy to identify. Nonetheless, the site of the ambush and massacre of Varus' legions in AD 9 has now been located with reasonable certainty near Osnabrück in northern Germany, where archaeologists have recovered more than 3,000 military items, ranging from fragments of weapons and tools to clothing paraphernalia and personal ornaments.[78] And in the rather different setting of the seabed off western Sicily, underwater archaeologists have recently identified remains from the final naval battle of the First Punic War (241 BC), with the recovery of ship timbers, bronze helmets and, most remarkable of all, nearly a dozen inscribed and decorated bronze ships rams.[79]

[73] Overview in James 2013; Bishop & Coulston 2006 for military equipment; essays on visual (and literary) representations in Dillon & Welch 2006.

[74] For camps during the Republic, see Rosenstein 2012a: 99–103; for forts in the Principate, see James 2013: 104–8.

[75] For the changing character of Roman swords, see James 2011a; for weapons and battle tactics, see Coulston 2007.

[76] Coulston 1990.

[77] James 2004 for the equipment, James 2013: 94–100 for an overview of the site and structures.

[78] Wells 2003: 43–55.

[79] Tusa & Royal 2012, Prag 2014.

CHAPTER I

War and Peace

The Roman state's record of war-making and overall military success across many centuries has understandably led to Roman society being seen as fundamentally militaristic. This chapter assesses this image by examining Roman attitudes and ideology, especially as reflected in religious ritual. It begins by investigating Roman attitudes to war as they developed through the main phases of Roman history, complementing this with consideration of attitudes to peace. The second section explores the related question of how Romans celebrated victory, together with the equally important question of how they dealt with defeat. Consideration of these subjects can provide insights into some of the fundamental assumptions which underlay the relationship between the Roman state and war, and the extent to which there were changes in those assumptions across the centuries.

1.1 Attitudes to War and Peace

It has become a truism in recent scholarship that Republican Rome was a militaristic state that waged war on an almost continuous basis, implying a positive view of warfare, at least on the part of the Roman elite. This section analyses the basis for these claims, draws attention to some important qualifications, and examines the extent to which the factors operating during the Republican period continued to do so in the later phases of Roman history. Discussion of these subjects also raises the complementary question of Roman attitudes to peace, although this is a subject that has received rather less attention in modern scholarship.

Historiographically, an emphasis on the militaristic features of the Roman Republic emerged in the late 1970s, perhaps influenced in part by contemporary disenchantment with modern imperialist ventures in the aftermath of the Vietnam War.[1] It certainly represented a reaction against the longstanding view of Rome as a reluctant imperialist. The previous influence of this older paradigm, with its benign view of Roman aims, can be accounted for in a variety of ways,[2] but an important element was acceptance at face value of Roman accounts of the so-called fetial law. These accounts, above all those of Livy (1.32) and Dionysius of Halicarnassus (2.72), describe the activities of the fetial priests, who undertook various functions relating to Rome's dealings with other peoples and states. In addition to the formalisation of treaties and the surrender of treaty-breaking Romans to an enemy, their functions were said to have included the performance of rituals designed to ensure that Rome only ever waged wars that had divine approval. These rituals involved the fetial priests seeking redress for injuries by travelling to the relevant frontier and invoking Jupiter as witness to the justice of Roman demands. Such practices seemed to lend credence to the idea that Rome had a well-established 'just war' tradition and could not therefore have engaged in territorial expansion without good reason. However, more recent scholarship has raised doubts about placing too much weight on the significance of the fetial law. While such rituals are plausible in the context of Rome's very early history, when it was one among a number of communities competing for position in Latium and was having to deal with the fallout from localised activities such as cattle-raiding, they become increasingly less so as Roman warfare extended through the Italian peninsula and into the wider Mediterranean world.[3]

While the revisionists of the late 1970s expressed scepticism about the relevance of the fetial law, the argument for a militaristic Rome was based primarily on renewed emphasis on other features of the Roman state and society which, while well-known, had been underplayed in this context. For one thing, war was clearly an important consideration in the organisation of Republican political institutions. There was a strong link between citizenship and military service, reflected in the oldest gathering of Roman

[1] The revisionist view was argued independently by Crawford (1978: 51–3), Hopkins (1978: 25–47) and (in most detail) Harris 1979, esp. ch. 1 (with a recent restatement of Rome's unusual aggression in Harris 2016: 37–43). For the possible influence of the contemporary context, see Rich 1993: 41–2.

[2] See Linderski 1984 for the historical contexts of some of its leading proponents (Mommsen, Holleaux, Frank).

[3] For detailed discussion and references, see Rich 2011. For ethical/philosophical reflections on 'just war' in the late Republic/early Principate, see Chlup 2014.

citizens, the centuriate assembly, being arranged into voting units according to the relative ability of individual citizens to provide their own arms (see further in Section 2.1), while the primary role of the city's senior magistrates, the annually elected consuls, was to lead the Republic's armies on campaign. Military success in this role was regarded by members of the elite as the best way to achieve renown for oneself and one's family, a principle confirmed by the ceremonial occasion of the triumph, when a victorious general paraded through the streets of the city to public acclaim (see further in Section 1.2). The attitude of the rank and file to war is less easy to determine with certainty because the ancient sources were written by and so reflect the views of the elite, although it has been argued that the prospect of booty is likely to have encouraged a favourable view.[4] Rome's treatment of defeated communities in Italy also implies a predisposition to war: rather than requiring payment of tribute, Rome stipulated provision of troops – 'taxing military labour instead of material resources'[5] – and for Rome to benefit from that provision presupposed that it would engage in further warfare: 'Wars were the very essence of the Roman organisation.'[6]

During the middle Republic Roman forces seem to have found themselves engaged in campaigns on an almost annual basis, corroborating the idea that the elite was hardly reluctant to engage in warfare. The continuous nature of warfare during much of the Republican period has seemed to find further support in the ritual associated with the temple of Janus, in the forum, whose doors were apparently only closed when Rome was not at war. A number of ancient sources claim that, prior to the reign of Augustus, they were closed on only one occasion, namely after the conclusion of the First Punic War in the mid-third century.[7] Ancient religious rituals associated with Mars, god of war, which cluster in March and October – the start and finish of the campaigning season, at least during the early centuries of the Republic – have also seemed to reinforce the idea that war-making was a fundamental feature of the city's annual cycle.[8]

There is no denying the implications of many of these features of Roman practices, or the essential validity of the claim that the elite of Republican Rome was favourably disposed towards making war. Some qualifications may, nonetheless, be noted. First, there is the question of the regularity of war-making. While there were probably few years when the

[4] Hopkins 1978: 38, Harris 1979: 102–3. [5] Scheidel 2008: 39.
[6] Momigliano 1975: 46; cf. Crawford 1978: 53. For reservations about Momigliano's view, see Harris 1984b.
[7] Varro *Ling.* 5.165, Livy 1.19.3–4. [8] Beard *et al.* 1998: vol. 1, 43.

Republic was not engaged in warfare somewhere, the level of commitment of military resources could vary quite significantly; there were years when campaigning was very limited. More specifically, there is good reason to think that much of the fifth century BC was relatively peaceful,[9] while from the mid second century BC onwards to the end of the Republic '[external] warfare became intermittent, and wide fluctuations can be observed between periods of intense fighting and interludes of relative calm'.[10] Indeed, Polybius claims that the senate decided to initiate war against the Dalmatians in 156 BC because, among other considerations, 'they did not wish the men of Italy to become weak and womanly in any way because of the long period of peace – for it was now twelve years since the war with Perseus and the campaigns in Macedonia' (32.13.6–7).

Secondly, reservations have been expressed about the significance of the rituals associated with Janus and Mars,[11] while another aspect of Roman religious practice warrants attention, namely the physical separation of military and civilian activities by the *pomerium*, the sacred boundary of the city. It is Roman acknowledgement of the *pomerium* that accounts for the centuriate assembly – the citizen body at arms – meeting on the Campus Martius (the 'Field of Mars', which lay outside the *pomerium*), for the temple of the goddess of war, Bellona, being erected in this same area in the early third century BC, and for triumphing generals needing to seek special dispensation for their troops to enter the city.[12] The maintenance of this separation during the Republic implies important limits to the militarisation of Rome.

A final important qualification from a rather different angle has been provided by a more recent study of Roman imperialism during the fourth and third centuries, in which Arthur Eckstein has argued that Rome was not unusual in being predisposed towards war. Eckstein situates Rome of the fourth and third centuries BC in the wider context of, first, an Italian peninsula, and then, a Mediterranean world, which, drawing on modern international relations theory, he characterises as 'anarchic' because no one state was in a position of dominance and there was no established framework for dealing with interstate disputes. This 'multipolar' world engendered an environment in which states had to organise for war or succumb to their neighbours.[13] Rome's militarism is not in doubt, then, but it was not exceptional. If that claim is accepted, then it implies a somewhat different view of the character of Roman imperialism – one in which the

9 Rich 2007a: 11–13.
10 Cornell 1993: 157; cf. Rich 1993: 44–9.
11 Rich 2013: 543–5.
12 Rüpke 1990: 29–41.
13 Eckstein 2006.

overriding emphasis is not just on Roman bellicosity and aggression, but in which allowance is also made for Rome's need to respond to the bellicosity and aggression of its neighbours.[14] In a sense, this picture represents a partial shift back towards the older paradigm, albeit via use of modern political science theory, without abandoning the insights gained from the revisionist arguments of the late 1970s. It may, however, still not give sufficient weight to the implications of Rome's arrangements with its allies. While Eckstein recognises the importance of Roman manpower resources in accounting for Rome's rise, his study does not perhaps give sufficient weight to what these arrangements imply for its incentive to initiate war.

The second century saw a dramatic change in the Mediterranean world, whereby the configuration of states shifted from multipolar competition towards unipolar dominance by Rome. This no doubt helps account for the development already noted, that Rome waged external wars less continuously during the final century or so of the Republic. At the same time, the fact that Rome nonetheless continued to wage expansionist wars even after the 'anarchic' context of the fourth and third centuries had resolved to a more orderly situation suggests that Roman militarism was not just a response to the militarism of other states. This was also the context in which civil war first emerged as a significant phenomenon in Roman history, starting with the so-called 'Social War' against Rome's Italian allies in 91–89, followed rapidly by the conflicts associated with Sulla in the 80s and Sertorius in the 70s, and then the civil wars of the 40s and 30s which brought the Republic to an end. This phenomenon partly reflected the fact that, with the further expansion of Rome's territorial empire, the political and material stakes had become even higher. But it was surely also related to the fact that Rome was no longer competing for survival against other states in the multipolar world of the fourth and third centuries BC.

Augustus' new regime signalled an important departure in relation to one of the most significant factors underlying the positive view of war which prevailed during the Republic – namely, the elite's pursuit of military glory. His constitutional control of most of the provinces where legions were stationed from 27 BC onwards meant that the governors of those provinces were his legates and therefore any victories they won did not formally entitle them to a triumph. As a result, the privilege of holding a triumph quickly became restricted to the emperor, or occasionally members of the imperial family who might hold a command in their

[14] Cf. Chaniotis 2005: ch. 4 for the aggression of Hellenistic kings.

own right, with the last recorded triumph by a member of the senatorial elite being held in 19 BC.[15] Instead, successful governors acting on the emperor's behalf had to make do with certain symbolic tokens, such as the right to wear a laurel crown at the games and the award of a statue.[16] There is debate as to whether or not this was an intentional by-product of Augustus' re-organisation of the provinces in 27 BC,[17] but whatever the answer it certainly suited him that members of the elite could no longer gain the kudos arising from a triumph, which might otherwise form the basis of a challenge to his authority. Even then, successful generals were often viewed with suspicion by emperors who lacked military experience, most famously Domitius Corbulo during the reign of Nero, but also Agricola under Domitian, and Salvius Julianus during the reign of Commodus (Cass. Dio 73.5.1–2). This political concern may have encouraged emperors to adopt a more cautious attitude towards allowing subordinates to engage in further territorial expansion and so helps to explain this distinctive feature of imperial policy during the Principate.[18]

This trend towards restricting the opportunities of the senatorial elite to celebrate military achievement led ultimately to their effective exclusion from holding military commands from the later third century onwards. Whether or not this was due to a specific measure on the part of the emperor Gallienus as some sources claim, the practical pattern is clear, with military commands now the monopoly of members of the equestrian order with military experience.[19] While the equestrian order had not traditionally been so far removed from the senatorial elite in social prestige, this third-century development nonetheless marked a significant change, since, in contrast with the situation in earlier centuries, many of these equestrians were men who had achieved this status through military service, often rising through the ranks.[20] The culmination of this development was the emergence during the final decades of the third century of emperors from this background, many of them with origins in the Balkans, which had become a major recruiting region by the third century.

[15] Lange 2016 highlights late Republican precedents for Augustan changes.

[16] Such 'triumphal ornaments' were still valued by their recipients: see, e.g., the famous inscription commemorating the career of Plautius Silvanus from the mid-first century: *ILS* 986.

[17] Cf. Beard 2007: 68–71, Rich 2013: 556.

[18] Cf. Cornell 1993: 162–4. Sidebottom 2005 suggests that emperors did not need to add territory because they accrued enough resources through inheritance and confiscation of property from the elite, whereas Harris (2016: ch. 4) emphasises financial limitations as a constraint on military campaigns.

[19] Davenport 2018: 485–7, 534–9.

[20] Davenport 2018: 509–20.

These important changes reflected the altered geopolitical circumstances of the Roman world, which experienced significant military setbacks and uncertainty during the mid-third century as it came to terms with the emergence of the Sasanian Persian regime to the east and of more powerful barbarian groupings to the north. Civil war also once again became a significant form of conflict. One way of viewing these developments is as the reverse process to that experienced during the Republic, with movement away from a unipolar world dominated by Rome to a multipolar one in which the Roman empire, while still a major player, was once more having to compete against other powerful players. The resultant emergence of 'soldier emperors' and expanded military forces has often been characterised as marking a process of militarisation, but if so, then once again some important qualifications need to be registered.

First, although the rhetoric of imperial victory remained unchanged (see further in Section 1.2), the Roman state in this period was rarely in a position to undertake imperial expansion, which is an obvious, but important, difference from the militarism of the Republican period. Secondly, while emperors of the later third and fourth centuries typically came from military backgrounds, the re-organised state which Diocletian and Constantine put in place was one in which there was a much sharper separation between military and civilian roles than had been the case during the Principate; the rationale for this may have been primarily a concern to place limits on the powers of those holding military commands, but it nonetheless represents a significant limit to militarisation. Thirdly, from the end of the fourth century onwards, not only was it the case that it became rarer for emperors to have a military background, but they almost never led their armies in person. This important shift was partly the result of the re-emergence of the dynastic imperative and partly a reaction to the death in battle of two emperors in the 360s and 370s. Fourth-century military emperors wished to consolidate their legacies through establishing a dynasty, and when the last of these emperors, Theodosius I, died prematurely while still in his 40s, he was succeeded by relatively young and militarily inexperienced sons.[21] This new pattern did, however, create opportunities for ambitious generals to exercise political power at the imperial court, which those in the west proved particularly adept at taking.[22]

It is only in the sixth century that it is possible to talk once again about significant Roman imperial expansion, in the context of the emperor Justinian's campaigns to remove the Vandal regime in north Africa and

[21] Lee 2007: 30–37. [22] Lee 2013a: ch. 5, McEvoy 2013: chs. 6, 9.

the Gothic regime in Italy. Strictly speaking, however, these were cases of the eastern half of the empire regaining territories lost during the fifth century, while Justinian's decision to undertake these campaigns appears to have included a strong element of religious justification, in so far as the Vandal and Gothic regimes supported the heterodox Arian form of Christianity.[23] Moreover, despite these campaigns achieving some success, it remained the case that Sasanian Persia was an established fixture in the wider geopolitical scene as a power of comparable resources and influence to the Roman empire, which acted as a significant limit on the ambitions of Roman emperors. Although that limit was in principle removed when the Islamic invasions of the early seventh century overthrew the Sasanian regime, the Roman empire also lost much of its eastern territory, including the economically critical region of Egypt, forcing a radical re-evaluation of state organisation and priorities.

Given the prominence of war during the Republic, it is unsurprising that the subject of peace was much less developed as a feature of Roman discourse and ritual: 'republican Latin is rich in words pertaining to war, poor in praises of peace', and Pax as a personified deity does not make an appearance until the end of the Republic.[24] The Principate, on the other hand, is traditionally associated with the phrase *pax Romana*, as in Pliny the Elder's famous dictum about 'the immeasurable majesty of the Roman peace' (*HN* 27.3).[25] Augustus promoted peace as an ideal, most visibly through his Altar of Augustan Peace, but also in a range of other media.[26] This was partly about advertising his claim to have ended civil wars (*RG* 34.1), which had had such a negative impact on communities around the Mediterranean. Their appreciation of internal peace was reflected in dedications to Augustus in regions not usually exposed to war,[27] and it was a theme endorsed by other emperors in the first century – in a senatorial decree under the emperor Tiberius referring to 'all the evils of civil war [which] had long since been laid to rest through the divine will of Augustus',[28] and in Vespasian's construction of a Temple of Pax in Rome after the civil war of 68–9.[29] However, Augustus' advertising of peace also

[23] Lee 2013a: 259.
[24] Quotation: Linderski 1984: 152. Republican *Pax*: Weinstock 1960: 44–52, Cornwell 2017: ch. 1.
[25] General discussions of the subject include Woolf 1993, Hardwick 2000, Rosenstein 2007, Cornwell 2017: 187–200.
[26] Weinstock 1960: 47–50, Galinsky 1996: ch. 4, Cornwell 2017: ch. 5.
[27] *SEG* 4.490 ('a saviour who brought war to an end': Asia, 9 BC), *ILS* 3786, 3789 (dedications to Augustan peace: Baetica, Spain and Narbo, S. Gaul).
[28] *SC Cn. Piso*, ll. 46–7 (AD 20) (= Eck *et al.* 1996: 42).
[29] Cornwell 2017: 191–4.

reminded Romans of the commitment he made in 27 BC to the pacification of those provinces which were insecure either because of the risk of internal rebellion or from external threats – a commitment crucial to Augustus' political supremacy since it justified his control of the empire's armed forces.[30]

As implied by the term 'pacification', peace was not, in Roman thinking, the outcome of mutual agreement between equals based on principles of justice, but rather the result of others submitting to Roman rule. This is reflected in a wide range of sources, such as the following: when dealing with a north Italian people in 181 BC, a Roman commander is said to have told their envoys that he only made peace with people who had surrendered (Livy 40.25); in his *Res Gestae*, Augustus referred to 'peace achieved through victories' (13); and another Roman commander is presented in 69 as reminding an assembly of Gauls that peace between peoples can only be maintained by arms (Tac. *Hist.* 4.74.1). In a similar vein are the sentiments attributed to non-Roman leaders, famously equating Roman peace with slavery and destruction (Tac. *Hist.* 4.17, *Agr.* 30). 'In Rome even peace was aggressive.'[31]

Indeed, there was a tradition that viewed peace in negative terms, maintaining that the lack of an enemy to fear encouraged a relaxation of moral virtues and a consequent weakening of the state. This was the attitude that underpinned the senate's decision to initiate war against the Dalmatians in 156 BC (Polyb. 32.13.6–7); it was also Sallust's diagnosis of the development of factional violence in late Republican politics following the destruction of Carthage in 146 BC (*Iug.* 41, *Cat.* 10), echoed in Tacitus' view that the Syrian legions in the mid first century were 'sluggish from a long period of peace' (*Ann.* 13.35.1) and that the Britons were more warlike than the Gauls because 'long years of peace had not yet weakened them' (*Agr.* 11.4).[32]

The view that the Roman empire should only make peace from a position of dominance persisted through the Principate and into Late Antiquity, even when circumstances forced the empire to engage in negotiation. In 89 the emperor Domitian agreed a peace with the Dacian king Decebalus, whose terms apparently included the empire giving Decebalus large sums of money on a regular basis (Cass. Dio 67.7.4) – a step which subsequently provided ammunition for critics (Plin. *Pan.* 12.2). Likewise during Late Antiquity the empire effectively bought peace on

[30] Rich 2012a. [31] Linderski 1984: 152.
[32] Mattern 1999: 205–6 for further examples; Wheeler 1996 for the *topos* of lax Syrian legions.

many occasions through the provision of financial subsidies to neighbouring states and groups.[33] Despite the practical benefits of such a pragmatic approach, the policy continued to draw opprobrium from members of the elite: in the words of one senator reacting to the proposal to give Alaric the Goth money not to enter Italy in the early fifth century, 'This is not peace, but slavery!' (Zos. 5.29.9).

Given the generally positive attitude to war throughout Roman history, and the limited character of discourse about peace, it is perhaps surprising to find even the occasional reference in Roman sources countenancing the possibility of an end to war. In phraseology reminiscent of Old Testament prophecies (Is. 2.4 , Mic. 4.3), the poet Martial represents a scythe (*falx*) as commenting that 'the settled peace of our emperor has bent me to unwarlike uses; now I belong to the farmer, where previously I was the soldier's' (14.34); the context, however, appears to have been Domitian's victory over the Chatti in the early 80s, which makes the sentiment less significant than at first sight.[34] In 382 the philosopher and orator Themistius commented in a speech delivered before the emperor Theodosius I that he had heard that those living in Thrace – the scene of much recent bloodshed between Romans and Goths – 'are now turning the metal of their swords and breastplates into hoes and pruning hooks' (*Or.* 16.211B); but again, the context makes this less significant – Theodosius' need to justify his reaching an accommodation with the Goths, rather than driving them out of the empire.[35] Similarly, in the mid-sixth century a Roman diplomat can be found extolling peace as 'very clearly a good thing for all mankind', in contrast to 'the uncertainties of war' (Men. Prot. *fr.*6,1 ll.34–6), but this was an argument from necessity, as the empire sought, from a position of relative weakness, to persuade Persia to accept its proposals.

More intriguing is the claim that the third-century emperor Probus planned to abolish the armed forces. The fullest statement of this appears in the *Historia Augusta*'s biography of Probus (20.3–6; cf. 23.2–5), but this is a particularly problematic source, and its reputation as a sophisticated literary jest from the late fourth century has led one commentator to interpret this passage as a knowing parody of Old Testament prophecies and their Christian interpretation.[36] The appearance of the same story in abbreviated form in other sources (Aur. Vic. *Caes.* 37.3, Eutr. 9.17.3) has prompted the alternative view that it is simply voicing a desire to see the restoration of senatorial authority at the expense of the military.[37] Neither

[33] Lee 2007: 119–22. [34] Weinstock 1960: 52, Leary 1996: 10, 88.
[35] Heather 1991: 157–81. [36] Paschoud 2001: 146–61. [37] Rüpke 1990: 21.

interpretation leaves any scope for the story to express genuine anti-war sentiments.

The suggestion that it is a parody of Christian views does, however, serve as a reminder of a potentially important new factor for attitudes to war and peace in Late Antiquity. While Christian teaching, as reflected in the Bible and in the writing of the early church fathers, was by no means consistent or unequivocal in its opposition to war, it did nevertheless include a greater willingness to question the use of violence, while, building on Old Testament prophecies, it also offered a more positive view of peace. This did not, however, translate into any re-evaluation of government priorities with the advent of Christian emperors from the fourth century onwards. Constantine and his more immediate successors were military men by background who faced serious external and internal military challenges and who did not have the time or inclination to consider the implications of their espousal of Christianity for war and peace beyond a traditional mindset that saw the Christian God as a potentially surer guarantor of military success. The thinking of the Christian bishop and intellectual Augustine about the issue of war and justice, while important in the longer term, did not have any immediate impact on elite attitudes, and the most notable trend during the final centuries of antiquity was an increasing tendency to view war in terms of religious conflict, whether against Zoroastrian Persia or heterodox Arian Vandals and Goths.[38]

1.2 Celebrating Victory, Dealing with Defeat

Given Rome's overall record of military success during the Republican period, it is unsurprising that there developed a range of rituals associated with the celebration of victory. These rituals illuminate the close interrelationship between war and religion in Roman culture, while also highlighting the political implications of military success. This section outlines these features during the Republic, while also charting their evolution in subsequent phases of Roman history. Even during its periods of greatest military success, however, Rome experienced temporary setbacks and defeats. Despite the popular image of Roman invincibility, its military power did have its limits, even during the Republic and increasingly so during Late Antiquity. So alongside a consideration of changing patterns in the Roman celebration of victory, it is also instructive to

[38] Discussions of this large subject include Noethlichs 2001, Lee 2007: ch. 7, Swift 2007.

examine how the Romans responded to defeat: how was it explained and did those explanations change over time?

The natural focus of some of the rituals associated with military success during the Republic was the divine personification of that success, the goddess Victoria. Although there is, inevitably, debate about the origins of the cult, her presence was manifesting itself in the city through a variety of media from the early years of the third century BC, when a statue of Victoria was said to have been put up in the forum and a temple dedicated to the goddess was constructed on the Palatine Hill. Since the latter was done prior to a campaign during the Third Samnite War, it was evidently a strategy for achieving divine support, as also was Cato the Elder's vow, and subsequent construction, of a temple to Victoria Virgo while campaigning in Spain in the 190s. Victoria also featured on coinage from the third century onwards, while other deities acquired epithets associated with victory during this period, above all Jupiter Victor and Mars Invictus.[39] In a more immediately military context, winged Victory motifs also adorned some of the ships rams recently recovered from the site of the final naval battle of the First Punic War in 241 BC, while Roman Victoria received honour in a Samnite ritual context in the second century BC.[40] The cult of Victoria received further elaboration and prominence in the increasingly competitive atmosphere of late Republican politics, as Marius set up many statues in her honour following his successes against the Cimbri and Teutones, only to be outdone by Sulla who, in addition to statues and trophies, established games in her honour. Pompey in turn built temples for Venus Victrix, Hercules Invictus and Minerva Victrix, while Caesar expanded Sulla's games.[41] Finally, Augustus made Victoria an integral element of state procedure by placing a statue of the deity in the senate house, as well as an altar dedicated to her, on which senators offered incense at the start of meetings.[42]

An important, more general way in which military success was celebrated during the Republic was through the granting of *supplicationes* – days of thanksgiving to the gods following news of a significant military victory. Requiring the senate's sanction, these occasions involved the offering of prayers accompanied by sacrifices, typically over a period of one to five days during the third and second centuries.[43] Once again, late

[39] Weinstock 1957: 215–18. [40] Rams: Tusa & Royal 2012: 42–3 Samnites: Dench 2018: 130.
[41] Weinstock 1957: 224–9, 233.
[42] Pohlsander 1969, Cornwell 2017: 98–102. For the iconography of Victoria during the Republic and beyond, see Hölscher 1967.
[43] E.g., Livy 45.2 (Pydna, 168); general discussion: Halkin 1953.

Republican political rivalries brought inflationary pressures to bear on the number of days of thanksgiving granted, offering as it did an obvious way of calibrating the significance of a victory against others. Pompey's achievements during the 60s were considered to warrant ten days, but Caesar's were then awarded fifteen days in 57, then twenty days twice in the late 50s, forty days in 46, and then fifty in 45.[44] The culmination of this trend was Augustus' claim to have been awarded a total of 890 days of supplications during his life (*RG* 4.2).

The grant of a *supplicatio* during the Republic was almost always the first step in the process leading to the best known and most important ritual of victory – the triumph.[45] The granting of a triumph by the senate allowed a victorious commander to parade through the thronged streets of Rome in special regalia and a four-horse chariot, preceded by his troops, wagonloads of booty, prisoners and, sometimes, pictorial displays of battle and of conquered cities or peoples – 'a spectacle in which generals bring right before the eyes of their fellow-citizens a vivid impression of their achievements' (Polyb. 6.15.8). The event culminated in the commander ascending the Capitol and offering sacrifice in the temple of Jupiter – the clearest reminder of the religious dimension of the occasion. However, it was also an event that could not fail to enhance the renown of the commander and strengthen his political influence and that of his family, and as such was eagerly sought after by any ambitious member of the elite. The senate's permission to stage the ceremony was required and although there has been much debate about the specifics of any formal requirements, the need to obtain permission served to enhance the event's prestige.[46] The distinction and prestige of the triumph was further enhanced by the existence of the lesser form of celebration known as an *ovatio*, in which the commander progressed on foot, rather than in a chariot, and wore the normal dress of a magistrate, rather than triumphal regalia, and a crown of myrtle, rather than laurel.

One can easily imagine the attractions of the public acclaim that such an event potentially offered, and yet a triumph lasted only a day, or very occasionally two or three days, while some were very routine affairs. However, there was a range of strategies for ensuring that the occasion

[44] Halkin 1953.

[45] Beard 2007, with succinct overview in Rich 2013: 551–7 (including caveats about some of Beard's conclusions); Pittenger 2008 discusses senatorial debates, and Östenberg 2009 the composition of the procession.

[46] Debate about requirements: Beard 2007: ch. 6, Rich 2014.

endured in public memory.[47] The most obvious strategy was to leave permanent memorials in public places, whether that be a statue, temple, portico or triumphal arch – all, of course, adorned with an appropriate inscription.[48] The entrance to the commander's house would also be decorated with captured booty as a reminder to visitors and passers-by, with subsequent owners apparently obliged to retain these features so that the house 'celebrated a triumph in perpetuity' and 'every day reproached an unwarlike occupant for entering someone else's triumph' (Plin. *HN* 35.7). During his lifetime the *triumphator* was entitled to wear a laurel crown at the games,[49] while after his death his family would perpetuate knowledge of his achievement by one of their number donning his triumphal robes and funeral mask at family funerals and by regular reference to his achievements in the funeral eulogies of other family members (Polyb. 6.53.6–7, 6.54.1–2).

As previously noted, Augustus' arrangement of the provinces effectively limited the holding of triumphs to members of the imperial family, thereby confirming the prestige associated with the occasion. However, during the following two centuries the frequency of triumphs decreased significantly. From an estimated average for much of the Republic of one triumph perhaps every one and a half years,[50] the period from Augustus to the early third century witnessed only thirteen triumphs by emperors, and three by imperial princes.[51] This change partly reflects the fact that only some emperors during the Principate participated personally in military campaigns and partly the fact of far fewer territorial additions to the empire, and it also helps to explain why the imperial *adventus* (ceremonial arrival in Rome or another city) assumed increasing importance over time, absorbing many of the features of a triumph.[52] These changes did not, however, reflect reduced interest in victory as a concept. If anything, the celebration of victory became even more prominent because the ideology of victory played such an important role in legitimating and underpinning the position and power of emperors – to the extent that some scholars have

47 Cf. Beard 2007: 18–31 for the specific case of Pompey; Popkin 2016 for the issue of memory.

48 Temples (of which approx. forty examples are known): Pietilä-Castrén 1987, Orlin 1997: 28–34, 199–202 (though temples could also be erected by generals who had not been awarded a triumph). Porticoes: Rich 2013: 555. Arches (of which six Republican cases are known): Kontokosta 2013. See also Popkin 2016: ch. 2 for triumphal structures from the period of the Punic wars.

49 Rich 2013: 555.

50 Pittenger 2008: 278. Rich (2014) emphasises the fluctuating frequency of triumphs across the Republican period.

51 Campbell 1984: 133–42, Rich 2013: 556–7.

52 MacCormack 1972: 725–6, Ando 2016: 413–14.

written about a 'theology of victory.'[53] That celebration took a range of forms. At the start of his reign Augustus made an emphatic statement about his victories at Actium (31) and Alexandria (30) by founding commemorative cities named Nicopolis ('Victory city') at each site, complete with regular celebratory games.[54] Subsequent successes by himself and his successors were memorialised in a variety of ways – through a proliferation of triumphal monuments, above all arches, trophies and columns (notably those of Trajan and Marcus Aurelius), not just in Rome but throughout the provinces; the increasing multiplication of *cognomina* derived from the names of defeated peoples as part of the imperial titulature (Germanicus, Dacicus, Parthicus and the like); intensification of victory as a theme on coinage; and regular religious festivals commemorating past victories.[55]

That emphasis on victory became even more important during the third century as, first, the empire found itself on the back foot militarily and then, secondly, the new style of Tetrarchic government that emerged towards the end of the century needed legitimation. When there were victories to celebrate, those opportunities were exploited. From Augsburg there is the recently recovered altar dedicated in 260 'to the sacred goddess Victoria' (who features in a relief on one side), commemorating the victory of a local commander over 'the barbarian Semnones or Juthungi' as they returned north from a raid into Italy, with the liberation of several thousands of prisoners.[56] Following his suppression of the breakaway Palmyrene state in the early 270s, the emperor Aurelian was able to celebrate a triumph in Rome in which the defeated Palmyrene queen Zenobia was paraded (thereby achieving what Augustus had failed to do vis-à-vis another eastern queen, Cleopatra),[57] while in 303, Diocletian and Maximian 'celebrated a triumph in Rome with notable pomp. Before their chariot went the wife, sisters and children of Narses, and all the booty, which they had plundered from the Parthians [i.e., Persians, defeated in 298]' (Jer. *Chron.* 227[m]). And because emperors now spent so little time in the city of Rome, such visits acquired added significance.

Indeed, during the fourth century the ideology of victory took on a new lease of life as it began to absorb Christian ideas linking the victorious emperor with a triumphant Christ, with the cross acquiring particular symbolic significance in this context.[58] At the same time the controversial

[53] Gagé 1933a, Fears 1981. [54] Lange 2009: ch. 4. [55] McCormick 1986: ch. 1.
[56] *AE* 1993.1231 with Bakker 1993 (image of Victoria relief at 371).
[57] Sources in Dodgeon & Lieu 1991: 105–7. [58] Gagé 1933b.

decision of Christian emperors in the later fourth century to remove from the senate house in Rome the altar of Victory which Augustus had placed there, and to resist attempts to have it restored, indicates an important divergence from Roman religious traditions.[59] By the sixth century, the traditional winged Victory which had featured on gold *solidi* for centuries was superseded by an angel holding a globe with a cross.[60] Another notable development was the way in which celebration of victory in civil war became more overt, perhaps because this was one category of warfare where emperors achieved unequivocal success in this period and they were able to brand their defeated opponents as usurpers.[61] The Arch of Constantine, with its inscription commemorating the suppression of 'a tyrant and all of his faction' (Maxentius), is the best known example of such celebration from the fourth century, but Constantius II also marked his defeat of Magnentius with an equestrian statue and obelisk in Rome, both accompanied by inscriptions referring to the elimination of 'the tyrant',[62] while Theodosius I erected an obelisk in Constantinople to mark his defeat of 'the tyrants' (Magnus Maximus and his son) (*CIL* 3.737). In the early fifth century a column with a spiral relief modelled on those of Trajan and Marcus Aurelius in Rome (but with the addition of some Christian symbolism) was erected in Constantinople to celebrate the defeat of Gainas in 400:[63] since Gainas was a general in the Roman military and had mobilised units against the emperor Arcadius, this was strictly speaking another instance of civil war, but as Gainas and many of his troops were of Gothic origin, the images may well have suggested success against a foreign enemy. A similar ambiguity may have been present in the emperor Anastasius' victory over the Isaurians in the final decade of the fifth century: while clearly a case of civil war, Isauria had long had a reputation as an untamed region within the empire.[64]

The emperor during Late Antiquity with the best grounds for celebrating victories of a more traditional kind was Justinian, above all following the overthrow of the Vandal regime in north Africa. During his reign Justinian advertised his successes through various media in Constantinople, including an equestrian statue in the square outside the imperial palace

59 Lee 2013a: 49–51. 60 Bellinger 1966: 36, 57, 67, 69, 133 etc., with Wright 1986: 77.

61 Cf. Wienand 2015.

62 *ILS* 731, 736. Constantius' famous entry to Rome in 357 was, according to one critical commentator, a triumph to celebrate this victory (Amm. Marc. 16.10).

63 Liebeschuetz 1990: 272–8.

64 Details of Anastasius' victory celebrations in McCormick 1986: 61, to which add *Anth. Pal.* 9.656 (erection of a palace commemorating the victory).

and a mosaic on the ceiling of the palace entrance showing him 'winning victories through his general Belisarius' and receiving booty and prisoners from vanquished Vandals and Goths (Proc. *Aed.* 1.2, 10). Most famous, however, was Justinian's staging of a triumph through the streets of Constantinople in 534 in which, unusually, the victorious general, Belisarius, was allowed to play a prominent part. Heralded as a revival of honours from 'olden times', Justinian nonetheless made sure that he was not upstaged by Belisarius, who was required to proceed on foot and, when he reached the hippodrome, to prostrate himself before the emperor alongside the defeated Vandal king Gelimer (Proc. *Bell.* 4.9). Needless to say, there was now no place for any culminating sacrifice to Jupiter.[65] Indeed in another triumph-like celebration towards the end of his reign, a focal point along the route of the procession was Justinian's visit to the church of the Holy Apostles.[66]

How, then, did Romans react when confronted by military failure? Although the Romans had a high win/loss ratio for much of their history, and even ordinary provincials in remote locations can be found asserting that 'the Romans always win',[67] they nevertheless experienced periodic defeats, even during times when the Roman state was predominantly militarily successful – as acknowledged by Lucilius, writing in the second century BC: 'the Roman people have often been beaten by force and overcome in many battles, but never in a whole war, in which lies all that is vital.'[68] There were of course a number of well-known military disasters during the Republic – the Gallic victory at the River Allia (390), which resulted in the sack of Rome itself, Hannibal's crushing victory at Cannae (216), the massacre of Roman troops by Germanic tribes at Arausio in southern Gaul (105), and Crassus' defeat by the Parthians at Carrhae (53). There were also some notable cases of Republican armies surrendering to the enemy in humiliating circumstances, such as to the Samnites at the Caudine Forks (321) and to the Numantines in Spain (137). But these are merely the most notorious from a much larger pool of Republican cases.[69]

65 Further discussion of this episode in McCormick 1986: 65–6, 125–9, Beard 2007: 318–28, Börm 2013.

66 Const. Porph. *De Cer.* Appendix (Reiske p. 497), with McCormick 1986: 67. The event celebrated the repulse of Cotrigur Huns from Constantinople in 559.

67 *AE* 1990.1016 = *SEG* 40 (1990), 1524 (Ḥisma, Arabia; mid-2nd c.?); the inscriber may have been an auxiliary soldier (Isaac 1998: 341).

68 613–14 M; cf. Livy 9.18.9, *Per.* 67 ('Romans cannot be conquered'). For this attitude in the context of the setbacks of the Hannibalic War, see Clark 2014: ch. 2.

69 Rosenstein 1990: Appendix 1 lists 102 defeats.

Instances from the Principate are fewer, but include the infamous massacre of Varus' three legions in Germany (AD 9), Caesennius Paetus' withdrawal from Armenia after the Parthians forced his surrender at Rhandeia (62), the defeat at Beth-horon suffered by Gaius Cestius at the hands of rebel Jewish forces (66), the destruction of a legion under the command of Oppius Sabinus by the Dacians (84/5), and the successive defeats of the governors of Cappadocia (Sedatius Severianus) and Syria (Attidius Cornelianus) by Parthian forces (161–2). From Late Antiquity, notable defeats include that of the emperor Valerian by the Persians (260), that of Julian by the same enemy a century later (363), the Goths' victory over Valens at Adrianople (378), and the failure of Leo's naval expedition against the Vandals in north Africa (468) – with many more, albeit less dramatic, possibilities available.

Given the strength of the Roman ideology of victory, it is worth considering how the Romans responded to military defeat. The question of response can be considered under two related headings – commemoration and explanation. With regard to commemoration, it has been noted that, unlike the Athenians, the Romans did not erect casualty lists or any other form of war memorial in their capital. However, this does not mean that the Romans were in denial about defeat: rather, 'they developed a different culture of commemoration, whereby Roman military disasters were incorporated into the state's religious calendar. Rome's response to heavy casualties in warfare was not to remember the individuals who had lost their lives, but to lament a serious reversal in Rome's fortunes and to seek to win back the gods' support.'[70] One well-known instance of this approach was the designation of 18 July as a *dies ater* ('a black day') on which no public business was to be conducted, following the defeat by the Gauls at the River Allia in 390 BC and the sacking of Rome – an anniversary that continued to be observed for many centuries well into the Principate. Likewise, a festival was designated for 8 June when the temple of *Mens* ('good sense') had been dedicated as a reminder of C. Flaminius' lack of good sense which had contributed to his defeat by Hannibal at Lake Trasimene in 217 BC.[71]

As this implies, when it came to explanations of defeat, commanders were often blamed, even if cases of actual prosecution were rare and reserved for commanders who appeared exceptionally culpable (e.g.,

[70] Cooley 2012b: 80. [71] Cooley 2012b: 83–4.

Claudius Pulcher after Drepana, 249, Servilius Caepio after Arausio, 105).[72] Soldiers could also sometimes be blamed and punished, although this was less common. Survivors from Cannae were effectively exiled to Sicily for not having stood their ground, while Roman setbacks at Numantia in Spain in the early 130s were attributed by some to the deficiencies of the troops, which Scipio Aemilianus had to put right before he was able to capture the city.[73] Interestingly, after the defeat by Pyrrhus at Heracleia in 280 both commander and soldiers were ordered to spend the winter in tents (Front. *Str.* 4.1.24).[74]

In some instances, however, a different kind of explanation was offered, involving contravention of religious ritual. The reasoning was that defeat was due, at least in part, to the Romans having alienated divine favour – a further illustration of the close relationship between warfare and religion in the Roman world. This explanation by no means always ruled out also laying the blame on the commander, as some of the following examples show.[75] According to some sources, Claudius Pulcher lost his battle in 249 because he ignored unfavourable auspices, C. Flaminius was said to have neglected a range of religious duties in Rome before leaving the city to meet his death at Trasimene in 217,[76] while Crassus famously ignored the report of adverse omens as he left Rome for his fatal Parthian campaign (Cic. *Div.* 1.29). The belief that disasters could be accounted for by alienation of divine favour persisted into the Principate. One of Augustus' responses to the news of the Varian disaster was to vow major games to Jupiter 'in the hope that the state might return to a better condition' – explicitly following, it is said, precedents set during the war against the Cimbri in the late first century BC and the Social War (Suet. *Aug.* 23.2). And Paetus is said to have advanced into Armenia in disregard of unfavourable omens – the horse carrying the consular insignia taking fright while crossing the Euphrates, and an animal due for sacrifice escaping outside the ramparts of the army's camp before its construction was complete (Tac. *Ann.* 15.7–8).

As for human responsibility, no instances of generals being prosecuted for defeat are known from the Principate, although that is partly due to the fact that a number of those defeated died in battle either by their own hand

[72] Rich 2012b, qualifying Rosenstein 1990 in important respects, while also conceding that some defeated commanders did enjoy subsequent electoral success.

[73] Rosenstein 1990: 102–3 (Cannae), 99–101 (Numantia).

[74] Further discussion in Rich 2012b: 88–94, arguing that the sources place more emphasis on commanders than Rosenstein 1990 allows.

[75] Emphasised by Rich 2012b: 96–8.

[76] References in Rosenstein 1990: 77–9.

(Varus, Severianus) or that of the enemy (Oppius Sabinus), while another died of natural causes soon after (Cestius Gallus). Of those who survived, Paetus is known to have been dismissed from his command (Cass. Dio 67.22.4), and it is likely that this was also the fate of Cornelianus, since a new occupant of his post was soon in place (*HA Verus* 9.2). However, blame was sometimes directed against the deceased, unable to rebut criticism and therefore obvious targets. There was a tradition in Roman historiography, represented above all by Velleius Paterculus, which portrayed Varus as militarily incompetent,[77] while Severianus' decision to advance into Armenia in 161 is presented by one source as having been influenced unduly by the charlatan oracle-monger Alexander of Abonutichus (Lucian *Alex.* 27). There is only limited evidence of blame being directed at the rank-and-file soldiers, although again the lack of detail in the surviving sources for many of these episodes may mask this. The primary example is the Syrian legions in the early 160s, whose defeats at the hands of Parthian forces are attributed in some sources to their discipline having been undermined by exposure to the luxurious lifestyle of eastern cities – a long-standing, but flawed, *topos* in Roman discourse.[78]

Crucially, none of these defeats occurred when the emperor was in direct command of Roman forces, thereby shielding the emperor from any direct blame. Although emperors did sometimes lead campaigns in person during the Principate (most obviously Trajan), there were a substantial number who did not, even when they had prior military experience, such as Tiberius and Hadrian. However, that changed during the third and fourth centuries as it became the norm for emperors to be militarily active, which also made them more vulnerable to criticism for defeat. Alongside that important change was a second: the growing prominence of Christianity gave the tradition of religious explanations a novel twist, as Christian writers offered their verdicts on the reasons for the empire's military setbacks. The most serious defeat during the third century was that of the emperor Valerian by the Persians in 260, because for the first time an emperor was captured by the enemy. Since Valerian had been responsible for initiating an empire-wide persecution of prominent Christians only a few years earlier, Christian writers interpreted his defeat and capture as the judgement of God, with one commentator in

[77] Velleius was, admittedly, a former soldier with knowledge of operations in Germany, but other sources present a more favourable view of Varus: see Wells 2003: ch. 5; Syme (1939: 511 n. 2) describes Varus as 'the official scapegoat'.

[78] Wheeler 1996.

particular highlighting the inversion of Roman traditions by describing Valerian as 'having been most deservedly triumphed over' (Lactant. *De mort. pers.* 5.4).[79] As for the two great military disasters of the fourth century, the emperor Julian's reversion to paganism offered a ready explanation to Christian observers for the debacle in Persia and his death. On the other hand, the emperor Valens was a Christian, but, even so, an explanation for his defeat and death at Adrianople was ready to hand in his support for heterodox Arian Christianity.[80]

Unsurprisingly, Christian explanations for the defeats and deaths of Julian and Valens did not go uncontested. Amongst the various claims as to who had struck the blow that ended Julian's life, the pagan rhetorician Libanius asserted that it was done by a Christian Roman soldier,[81] and Libanius later argued that the disaster at Adrianople showed that the gods were angry that Julian's death had not been avenged (*Or.* 24.16, 30, 40). Interestingly, the pagan historian Ammianus Marcellinus, although an admirer of Julian, nonetheless indicates at various points in his narrative that in invading Persia, Julian persisted in the face of inauspicious omens.[82] Other comments by Ammianus imply that some pagans attributed Valens' death to his persecution of pagan adherents during his reign, of which there had been a particularly vigorous episode in Antioch in the early 370s.[83] The famous request by the senator Symmachus to the emperor Valentinian II for the restoration of the altar of Victory to the senate house in 384, following its removal a few years earlier by the emperor Gratian, was supported in part by the argument that without maintenance of the cult of Victoria, and indeed pagan cult more generally, the empire could only expect further defeats at the hands of foreign enemies (*Relat.* 3.3, 9). There clearly developed a growing pagan conviction that it was the emperors' abandonment of paganism in favour of Christianity that was responsible for the empire's military setbacks, articulated most forcefully by the late fifth-century historian Zosimus who in turn reflected the views of the late fourth-century historian Eunapius.[84] It was such views, and especially the fallout from the Gothic sack of Rome in 410, that in turn prompted Augustine to embark on his monumental apologetic work *The City of God*, which aimed to rebut the notion that Christianity was responsible for the empire's decline, while his protégé

[79] Other accounts collated in Dodgeon & Lieu 1991: 58–65.
[80] Lenski 1997: 149–54.
[81] Lib. *Or.* 18.274–5 with Sozom. *Hist. eccl.* 6.2; cf. also Amm. Marc. 25.6.6 for early rumours of Roman responsibility.
[82] 23.5.4–13, with Liebeschuetz 1988, Matthews 1989: 126–7.
[83] Lenski 1997: 155.
[84] Treadgold 2007: 81–9, 107–14.

Orosius wrote a history which emphasised the severity of military (and other) disasters in the empire's pagan past compared with his own day.[85]

Explaining defeat in Late Antiquity was not, however, conducted solely at the level of religious causation. Human agency was also seen by many as playing a significant role. The sources for the events surrounding Valerian's defeat in 260 are very patchy, but it is evident that there were other explanations besides the Christian emphasis on divine judgement. Zosimus accounted for the debacle in terms of natural disaster and Persian treachery – that Valerian's army was severely weakened by an outbreak of plague, prompting him to seek a settlement with the Persian king Shapur I, who then perfidiously took Valerian prisoner during negotiations (1.36). However, this may be a case of one pagan defending another against criticism; other non-Christian sources refer to Valerian as being defeated in war by Shapur, though without blaming him for incompetence.[86] Because Julian was surrounded by so much religious controversy, it is less easy to discern whether ancient commentators viewed him or his troops as militarily responsible for the debacle. In the case of Valens, on the other hand, there are clear implications in contemporary sources that the emperor, some of his generals, and his troops were all targets of criticism for military incompetence from some quarters,[87] though since so many of them perished in the battle, such criticism was to a large extent academic.

After the death of Theodosius I in 395, it was rare for emperors to lead a military campaign in person, initially because Theodosius' sons were still relatively young, but probably also because of the desire to avoid the political instability which had ensued from the deaths in battle of Julian and Valens – and perhaps also recognition of the need to distance the office of emperor from direct blame for defeat. That blame was instead directed onto the relevant general. So, for example, when Belisarius suffered a defeat at Persian hands at Callinicum in 531, an official enquiry was held as a result of which he was dismissed from his post by Justinian.[88] Later in the century, during the reigns of the emperors Tiberius II and Maurice, a number of generals were dismissed following defeats.[89] Although these dismissals carried blame, they did not result in prosecutions, nor did they

[85] See further O'Daly 1999, Van Nuffelen 2012.

[86] Eutr. 9.7, *Epit. de Caes.* 32.5.

[87] Lenski 1997: 145–8.

[88] *Proc. Bell.* 1.18, Joh. Mal. 461–6, with discussion in Greatrex 1998: 200–7 – a case which illustrates how apportioning 'blame' is never simple, inasmuch as defeat, in any era, usually occasions mutual recriminations.

[89] Justinian: Theoph. Sim. 3.15.8, Joh. Eph. *Hist. eccl.* 6.10, Evag. *Hist. eccl.* 5.19. Philippicus: Theoph. Sim. 3.5.14–16. Petrus: Theoph. Sim. 7.2–5.

necessarily finish the prospects of the individual in question. Belisarius was able to resurrect his career through his role in saving Justinian during the Nika riot in Constantinople in 532, going on to lead the successful campaign against the Vandals the following year, while some of those in the later sixth century held further commands at later dates, not least because they were relatives of the emperor by blood (Petrus) or by marriage (Philippicus). The use of dismissal in this way suggests it served as a convenient strategy for deflecting blame from the person of the emperor.

CHAPTER 2

Military Service and Courage

Attitudes to war and peace, and the celebration of victory, were matters of ideology with potential relevance to all inhabitants of the empire, but there were also issues of ideology with a more specific bearing on those liable to military service and these are the focus of this chapter. First, there is the relationship between military service, citizenship and property ownership. The latter two had an established relationship with military service for much of the Republic, but over the course of time those relationships shifted, and tracing those shifts can be very revealing. Secondly, courage was understandably an important ideal for soldiers and their identity. How was it instilled and displayed, and how did its representation change over the course of Roman history?

2.1 Military Service, Citizenship and Property

From early in the Republican period, there was a close relationship between military service and Roman citizenship. One of the clearest indications was the character and organisation of the centuriate assembly – the citizen body responsible for the election of consuls and other senior magistrates, and for declarations of war – meetings of which were equated with the summoning of the citizen army.[1] Appropriately enough, the Campus Martius, the open space to the west of the city defined by the Tiber where the assembly met, was also the site for military training exercises and (by the first century BC) the levy,[2] while the assembly's organising principle – the distribution of the citizen body into 193 units known as centuries – defined both a citizen's military obligations and his voting power in that assembly.[3] In his famous analysis of Republican institutions, Polybius identified Rome's reliance on its citizens for military

[1] Varr. *Ling.* 6.93, Aul. Gell. 15.27. [2] Exercises: Rance 2000: 267 n. 104; levy: Varr. *Agr.* 3.2.4.
[3] For the functioning of the centuriate assembly, see Lintott 1999: 55–61.

service as an important factor in explaining its eventual defeat of the Carthaginians in the Punic Wars. The latter 'employ foreign and mercenary troops, whereas those of the Romans are citizens and natives of their own country . . . The result is that even if they happen to be defeated at the outset, the Romans carry on the war with all their resources, . . . knowing themselves to be fighting for their country and their children.'[4]

A natural corollary of the close relationship between military service and citizenship was the exclusion of slaves from service in the armed forces. Although explicit statements to this effect derive from the Principate (below), this was not a post-Republican innovation, but a long-established and fundamental principle. The clearest evidence is the exception to the rule. Following Hannibal's victory at Cannae, 'necessity and the shortage of freeborn citizens brought about another form of unprecedented levy: 8,000 young and physically fit slaves were armed, having first been asked individually whether they were willing to undertake military service and then purchased from their owners at public expense' (Livy 22.57.11). These so-called *volones* ('volunteers') served well for the next two years until, following victory in the battle of Beneventum in 214 and with the immediate manpower crisis easing, they were rewarded with their freedom.[5] However, the *volones* did not become a precedent for any further such officially sanctioned experiments during the Republic. Assumptions about manly courage being an attribute of the freeborn and about the inferiority of slaves were too ingrained.

If citizenship by birth was the *sine qua non* for military service, however, there remained a further requirement for entry to the legions, at least until the late second or first century BC – namely, ownership of property. Citizens were required to make a declaration of their property at the time of the census, usually held every five years (Figure 3). While all citizens were entitled to participate in meetings of the centuriate assembly, that assembly was organised on the basis that some citizens were more equal than others. Established as it was in a period before the state took responsibility for equipping troops, the assembly was arranged on the assumption that the amount of a citizen's property determined his ability to arm himself and hence the level of his contribution to the state's military needs. Hence the voting units (centuries) to which citizens were assigned were categorised on the basis of wealth, with those who contributed more

[4] 6.52.4–7. Polybius' observation here overlooks the important contribution of the non-citizen Italian allies, of which he shows himself well aware elsewhere in his history (see further Dench 2018: 125).
[5] Livy 24.14–16, with discussion in Welwei 1988: 5–17.

Figure 3 Left-hand end of the relief from the so-called Altar of Domitius Ahenobarbus, Rome, traditionally identified as a representation of the Roman Republican census (recently challenged by Maschek 2018); citizens make their declaration on the left, the armed figure at the right has been identified as the god Mars, while the figures in the centre show how soldiers were equipped in the late second century BC and reflect the close relationship between citizenship and military service during the Republic.
Louvre, Paris/Bridgeman Images

militarily also having a greater say politically when it came to voting in the assembly. The lowest ranking out of all the centuries was one comprising those citizens who were too poor to own any property of significance – the so-called *proletarii*, whose only contribution to the state was their *proles* (offspring) and who traditionally were not expected to serve in the legions.[6] They included the urban poor and stood in contrast to the *assidui* – those citizens who owned sufficient property and who were assigned to one of the remaining 192 centuries, depending on the amount of their property. Part of the rationale for this arrangement was the belief that those citizens who owned some property, even if relatively modest, had a stronger incentive to defend the Roman state at all costs (Aul. Gell. 16.10.6). It certainly contributed to an ideology that viewed the peasant farmer as the ideal candidate for military service, as typified by Cato the Elder's observation in the mid-second century BC that 'it is from the farmers that the bravest men and the sturdiest soldiers come' (*Agr.* pref.). Only in times of emergency might the *proletarii* be provided with arms and expected to fight:[7] otherwise, their military contribution

[6] Also referred to as *capite censi* ('those registered by their head') (Rich 1983: 290 n. 13). Precisely what the property threshold was (and hence the proportion of *proletarii* in the citizen body) is the subject of much debate: for references and discussion, see Rosenstein 2004: 185–8, de Ligt 2012: 98–105.

[7] Brunt 1987: 395. Rich 1983: 290–92 emphasises the limited evidence for this phenomenon, even during the Hannibalic War.

involved service as rowers in the fleet (Polyb. 6.19.3) – a role that, though dangerous in wars involving a significant naval dimension (e.g., the First Punic War), was viewed as significantly inferior in status to that of service in the legions, as reflected in freed slaves also serving as rowers.[8]

This close nexus between military service, citizenship and ownership of property was forged during the early and middle Republic, when Roman warfare mostly took place within or near the Italian peninsula. However, Roman overseas expansion, on the one hand, and civil war in the late Republic, on the other, brought pressures to bear on the links between these three factors. To take the latter first, heavy demands on manpower during the civil wars of the 40s and 30s induced Caesar, Pompey, Antony and Octavian to raise legions from non-citizens, thereby breaking the principle that legions were recruited only from Roman citizens and becoming a mechanism of enfranchisement.[9] As for the former, the changing relationship between military service and ownership of property is particularly complex and contentious. The championing of land reform by the Gracchi brothers and Marius' open recruitment of *proletarii* in the late second century have traditionally been seen as seminal episodes reflecting major changes in landowning in Italy with important ramifications for the character of military service. A number of influential studies, particularly during the 1960s and 1970s, developed the argument that Rome's long-term overseas military commitments in Spain and the eastern Mediterranean during the late third and second centuries made it increasingly difficult to sustain the original model of the peasant farmer who campaigned during the summer months before returning home to attend to his land. This development became intertwined with other consequences of Roman military success and expansion, above all the influx of overseas wealth which enabled members of the Roman elite to buy up land in Italy at the expense of small farmers, and the influx of large numbers of prisoners of war who could be deployed as slave labour to farm these expanding estates. The number of citizens who met the minimum requirement to be classified as *assidui* increasingly fell, even when the threshold was progressively lowered, creating a recruitment crisis which culminated in Marius' symbolic breaking of the link between landowning and military service.[10]

Offering as it did an elegant explanation for no less than the eventual fall of the Republic, this narrative proved compelling. It was, however, one

[8] Rich 1983: 290. [9] Haynes 2013: 37–8, Lavan 2019a: 33.
[10] Brunt 1971, Hopkins 1978: ch. 1.

based almost solely on the literary evidence, whose interpretation of the significance of the Gracchi and of Marius it took for granted. One of the crucial elements in more recent revisions of this traditional picture has been the introduction of the results of archaeological survey into the debate, since field surveys in parts of central Italy have demonstrated the continued existence of significant numbers of small farms to the end of the Republic and beyond. While further refinements in the analysis of the archaeological data have revealed a more complex and less straightforward scenario, the field survey results have nonetheless cast doubt on the idea of significant displacement of peasant farmers by large estates manned by slave labour.[11] Further modifications have also arisen, at least in part, from fresh examinations of epigraphic evidence. Recent research on the Roman family, particularly the age at which men married, has drawn on epigraphic data indicating that Roman men typically did not marry until their late 20s.[12] Although this data relates primarily to the post-Republican period, there is reason to think that Republican patterns were comparable, which is important because liability for military service began at age seventeen. It will therefore have tended to be the sons of small-holders on whom the main burden of military service fell, rather than the small-holders themselves, thereby reducing the impact of prolonged military service overseas on the well-being of Italy's small farms.[13]

Against this background, Marius' famous recruitment of *proletarii* for his African campaign in 107 assumes far less significance than often attributed to it. In the following decades troops continued to be drawn primarily from rural inhabitants rather than landless city-dwellers,[14] and no other general is reported recruiting *proletarii*, implying no shortage of *assidui*. In his account of Marius' action, the historian Sallust emphasises the eagerness of the *proletarii* to serve under him, in expectation of winning easy booty (*Iug.* 84.4), and it has been plausibly suggested that Marius enlisted them simply to capitalise on his popularity with the urban masses.[15] Rather than Marius' action, it is likely to have been the mass mobilisations associated with the Social War and the civil wars of the final decades of the Republic, when demand for manpower was at its height, which uncoupled the formal requirement of property ownership from service in the legions. A key factor in this was probably a change in the way the levy was conducted: instead of those eligible for service being

[11] Overviews in Rich 2007b: 158–60, de Ligt 2012: 179–82. [12] Saller 1994: 24–42.
[13] This, in greatly simplified form, is the argument of Rosenstein 2004: ch. 3.
[14] Brunt 1962: 74–5. [15] Rich 1983: 323–6; cf. Brunt 1987: 725, de Ligt 2012: 183–4.

required to present themselves for selection in Rome, *conquisitores* (lit. 'seekers out') were sent to localities to get recruits, presumably with no obligation to check whether they satisfied the property qualification.[16]

Despite this, the ideal of the farmer-soldier seems to have endured well beyond the Republic. Writing in the later second century, the rhetorician and philosopher Maximus of Tyre argued that 'if there is ever any fighting to be done, you will find in the farmer a soldier trained in genuine exertions' (*Or.* 24.6). Likewise, when considering in his military treatise whether recruits from the country or the city are more useful, Vegetius (writing in the late fourth or early fifth century, but also drawing on earlier sources) was in no doubt that 'the rural populace is better suited to arms' because of their acquaintance with hard work, their being content with little and their lack of exposure to luxury, and so 'from the country the main strength of the army should be supplied' (1.3). By contrast, he regarded many forms of urban employment (e.g., fishmongers, pastry-cooks, weavers) as wholly unsuitable preparation for military service because of their association with women.[17] To what extent this ideal was reflected in practice during the Principate is difficult to determine, with some evidence suggesting that there was a preference for recruits from social backgrounds involving small-scale property-owning, while other evidence points to uptake by the rural poor seeking a better life in the military.[18]

During Late Antiquity, 'recruits were levied on the same assessment as the land tax and the burden therefore fell exclusively on the rural population'.[19] Moreover, it was the under-urbanised Balkans and other similarly underdeveloped regions such as Isauria and Armenia that were regarded as the premiere regions for recruits, implying that soldiers were being drawn from struggling rural backgrounds. One of the most famous of these was a man who went on to become emperor in the early sixth century (Justin I): in the mid-fifth century 'three young Illyrian peasants named Zimarchos, Ditybistos and Justin set out to join the army because at home they had to struggle constantly against poverty and all its attendant hardships' (Proc. *Anecd.* 6.2). At the same time, there remained an appreciation of the need for training to turn recruits from a rural background into effective soldiers. In his account of the battle of Callinicum (531), Procopius observed that a unit from Lycaonia (central Anatolia)

[16] Brunt 1962: 74, Rich 1983: 327–30. [17] 1.7, with further discussion in Charles 2010.
[18] For differing views, see Carrié 1993: 107, Campbell 2002: 32–4. [19] Jones 1964: 615.

suffered heavy casualties at the hands of the Persians because 'they were very inexperienced, having only recently left off farming' (*Bell.* 1.18.39).

Once in the armed forces, however, these post-Republican recruits left their rural roots behind, at least until their discharge many years later, now that military service was a full-time occupation, usually involving a minimum of twenty years' commitment. Indeed third-century jurists indicate that soldiers were forbidden from buying land in the province where they were serving 'so that they may not be distracted from their military service by an interest in farming' (*Dig.* 49.16.13; cf. 9). This did not, however, prevent them from inheriting or renting land, which could be cultivated by relatives, slaves, hired labour or tenants, for which there is some documentary evidence from Late Antiquity.[20] Furthermore, there is one important category of late Roman soldier which has sometimes been regarded as a type of soldier-farmer – namely, the troops known as *limitanei*, who served in frontier provinces during the fourth to sixth centuries. Laws from the mid-fifth and sixth centuries refer to *limitanei* cultivating land and prohibit their land from being passed to others, which implies that the state recognised the role of such land in supporting *limitanei*. On the other hand, they continued to receive their normal pay, at least until the mid-sixth century. The characterisation of *limitanei* as farmer-soldiers has usually been made pejoratively in the context of critiques of the late Roman armed forces, with the implication that they had become no more than a part-time militia. Yet not only is there good evidence for the military capabilities of *limitanei* throughout these centuries, but the legions of the Republican period also show that farmer-soldiers could be a highly effective fighting force.[21] So while some late Roman soldiers may have become more involved in farming while still on active service, this should not automatically be read as symptomatic of a wider malaise.

As for the post-Republican relationship between legionary service and citizenship, it continued to be regarded as an important point of principle during the Principate. Documentary evidence relating to the formal process of enlistment includes the assertion by a recruit that he was 'freeborn and a Roman citizen and has the right to serve in a legion',[22] while there is a range of evidence for the prohibition of slaves from military service, including an explicit statement to this effect by an early third-century jurist

20 Rance 2018: 403–4.

21 Isaac 1988: 139–47, Whitby 1995: 110–16. The laws: *Nov. Theod.* 24.4 [443], *Cod. Iust.* 1.27.2.8 [534].

22 *CPL* 102, line 5 (Egypt, 92).

(*Dig.* 49.16.11) and a famous exchange of letters between the emperor Trajan and the younger Pliny in the early second century when the latter discovered two men of slave status in a group of provincial recruits; Trajan's advice that their action, if taken on their own initiative, warranted capital punishment leaves no doubt as to the continuing importance of the principle.[23] As during the Republic, so also during the Principate it is possible to find exceptions to this rule, but once again they are rare and feature in the context of crises – most famously after the massacre of Varus' three legions, when Augustus sanctioned the enlistment of slaves, albeit only after they had first been manumitted.[24]

During the Principate the Roman empire encompassed the whole Mediterranean basin, but although this meant that there was in theory a much larger potential pool of manpower available for recruitment, the continuing requirement of citizenship for service in the legions was a significant constraint, at least in the first century. Grants of citizenship to non-Italian provincials had increasingly been made during the late Republican period, especially in the west,[25] and it was descendants of these individuals who began to be recruited into the legions in growing numbers during the Principate, not least because citizens from Italy proved increasingly reluctant to undertake military service outside of Italy, involving as it usually did a posting to a distant frontier region on the Rhine, the Danube or in Syria. So while the principle of the citizen-soldier continued to be upheld in theory, those citizens serving in the legions increasingly came from different and diverse cultural backgrounds (see further in Section 3.1). In the eastern provinces, where individuals with Roman citizenship were less plentiful, legions gradually took to recruiting non-citizens who were then granted citizenship on enlistment.[26] Elsewhere, provincials who lacked citizenship could serve in the *auxilia*, which became a route by which provincials could acquire Roman citizenship, although not until the end of their period of service. The many surviving examples of military diplomas (Figure 2), which recorded these grants of citizenship, testify to the importance attached to this opportunity, by both the imperial authorities and the recipients, even if the number so enfranchised should not be over-estimated.[27]

[23] Plin. *Ep.* 10.29–30 (likely recruits for *auxilia*, rather than legions: Haynes 2013: 66, 71).
[24] Welwei 1988: 18–22. [25] Lavan 2019a: 29. [26] Brunt 1974: 95, Mann 1983: 41–9.
[27] Lavan 2019b argues persuasively that the total number of beneficiaries was significantly lower than previous estimates (fewer than 1.6 million across two centuries, compared with previous estimates of at least 3 million).

The emperor Caracalla's decision to extend citizenship to virtually all free inhabitants of the empire in 212 might appear to have marked the end of the citizen-soldier as a subject of any significance in Roman history.[28] Nonetheless, legal status continued to be relevant, in two respects. First, universal citizenship did not extend to slaves, and since slaves continued to be an important feature of late Roman society,[29] the legal status of soldiers continued to be a matter of concern. A series of laws from the fourth and sixth centuries reiterated the prohibition on slaves undertaking military service,[30] while other laws emphasised the importance of examining the freeborn legal status of would-be recruits (*Cod. Theod.* 7.2.1, 7.13.1). Needless to say, exceptional circumstances occasionally continued to prompt the suspension of this principle, notably during Radagaisus' invasion of Italy in 406, when the relevant edict made clear that the military service of the freeborn still remained preferable: 'although we believe that freeborn persons are inspired by love of homeland', nonetheless slaves were encouraged to enlist with the promise of their freedom and a generous gratuity of two *solidi*.[31]

Secondly, it is apparent that freeborn non-citizens continued to exist in the empire after 212, at least during the third century, despite Caracalla's universal grant – presumably reflecting the less than thorough approach to registration of new citizens by some imperial administrators.[32] In the mid-fourth century Julian referred to the presence of both foreigners and citizens in his army (*Ep. ad Ath.* 285B). Since Julian's forces are known to have included recruits from beyond the Rhine (Amm. Marc. 20.4.4), this is less surprising, and is certainly consistent with the wider evidence for emperors recruiting barbarians in significant numbers during the fourth century, even if traditional claims of 'barbarisation' are misconceived.[33] Some of these were from barbarian groups who had been defeated and been settled on land within the empire in return for providing recruits who, as *dediticii* ('those who have surrendered'), presumably retained non-citizen status, at least in the first generation. However, some were recruited from beyond the frontier or crossed the frontier of their own volition – and in some cases, achieved great success by rising to high rank. While the surviving sources provide no explicit discussion of the citizen status of such individuals, it is inconceivable that men of barbarian origin who held the

[28] For a re-assertion of the significant increase in citizen numbers that this decision entailed, see Lavan 2016.

[29] Harper 2011.

[30] Jones 1964: 1254 n. 14.

[31] *Cod. Theod.* 7.16.13, with further discussion in Welwei 1988: 167–80.

[32] Garnsey 2004: 143.

[33] For references and discussion, see Lee 2007: 83–5.

highest posts in the Roman military and were sometimes even rewarded with the consulship did not have citizenship. It therefore looks as if the distinction between citizens and freeborn non-citizens that persisted during the third century and into the fourth gradually disappeared, perhaps not because of barbarian incomers undergoing any formal process for becoming a Roman citizen (about which the sources are silent), but rather because citizenship had become 'a matter of participation and self-identification', above all through making use of Roman law in the practicalities of life.[34] Certainly, when detailing the ways in which a particular recruit satisfied enlistment requirements, a communication between two officers in the early sixth century did not include any reference to citizenship (*P. Ryl.* 609).

2.2 Courage as an Ideal

Polybius famously observed that 'courage (*andreia*) may be said to be the most important of all virtues in any state, but in none more than in Rome' (31.29.1). Earlier in his history, when describing the funeral ceremonies of the Republican elite with their elaboration of the deeds of the deceased and his forebears, he noted that 'the most important consequence of the ceremony is that it inspires young men to endure extremes of suffering for the common good in the hope of winning the glory that waits upon the brave' (6.54.3). As far as he was concerned, then, the ideal of courage was an important element in understanding Rome's military success during the Republic. This section aims to explore the role of this ideal in Roman warfare during the Republic and later periods, with a view to determining the extent to which it remained a constant or underwent change, as well as its potential to create tensions with other values relevant to Roman military performance.[35]

Republican Latin had a range of words to denote courage, including *audacia*, *fortitudo* and *animus*, but the term which was seen as best epitomising the ideal was *virtus*, not least because of its etymological roots in the word *vir*, 'man', in the exclusively masculine sense (as also the Greek *andreia*).[36] *Virtus* was closely related to military prowess in the thinking of Romans, reflected in and reinforced through the development of the honour shown to its divine personification, Virtus. An important stage

[34] Mathisen 2006, 2012. [35] Harris 2006: 300–7 sets out some of the issues.
[36] Discussion of terminology in McDonnell 2006: 59–62; Williams 2008, Gleason 2009 for reservations.

in the emergence of Virtus as an object of religious devotion was the vowing of a temple to Honos and Virtus in Rome by M. Claudius Marcellus in 222 BC. The context in which the vow was made leaves no doubt as to the military associations of the cult – Marcellus' defeat of Gauls at Clastidium in northern Italy, during which he famously killed the enemy leader and subsequently dedicated the latter's arms and armour as so-called *spolia opima*.[37] Likewise, Scipio Aemilianus dedicated a shrine to Virtus at the time of his triumph for his capture of Carthage in 146 BC, Marius built a temple to Honos and Virtus from the spoils of his victory against the Germanic Cimbri at Vercellae in northern Italy in 101 BC, while Pompey's theatre, erected to commemorate his eastern victories, included a shrine to Virtus.[38]

The importance of courage as part of the ethos of the Republican elite can be seen in such anecdotes as the elder Cato preferring, as a young adult, to enhance his reputation, not by promoting his successes in the law courts, but by advertising his achievements on the battlefield and the wounds he received on the front of his body; and the young Scipio Nasica leading a dangerous night-time mission to attack Macedonian forces from the rear in 168 BC, during which he despatched an enemy soldier with a spear in the chest – as he himself subsequently made a point of publicising.[39] It was of course possible for senatorial aristocrats to demonstrate courage when older and commanding armies, as the example of Marcellus taking on the Gallic leader at Clastidium shows, while the occasional deaths of Roman commanders in battle is a reminder of the dangers they confronted more generally.[40] However, as the case of Cato shows, it was in early adulthood, serving as military tribunes or in the cavalry, that the best opportunities for demonstrating personal courage presented themselves, above all in single combat.[41] When describing how elite funeral ceremonies inspired young men to acts of bravery, the first example Polybius gives is of how 'many Romans have volunteered to engage in single combat so as to decide a whole battle' (6.54.4), and there is sufficient evidence to suggest that such formalised single combats

37 Discussion in McDonnell 2006: 212–35, Clark 2007: 67–9. On the *spolia opima*, see Flower 2000, Rich 2013: 557–8, with references; an indication of their prestige is the report that Caesar desired the *spolia opima*: Cass. Dio 44.4.3.

38 McDonnell 2006: 239, 268–9, 275–80, 295, Clark 2007: 124–8, 223–5.

39 Cato: Plut. *Cat. Mai.* 1.5, with Leigh 1995 for further discussion of and references to honourable scars. Nasica: Plut. *Aem.* 15.3–16.2.

40 Rosenstein 1990: 117–21.

41 For Cato as military tribune: Nep. *Cato* 1.2, Plut. *Cat. Mai.* 3.3; for opportunities for cavalry, see McCall 2002: 83–99.

probably occurred on a regular basis throughout the Republican period. One member of the senatorial elite from the late third and early second century BC is reported as claiming that 'I have challenged and fought an enemy twenty-three times; in all cases it was I who carried off the spoils from hand-to-hand combat; I have a body marked by honourable scars, all incurred with my face to the enemy' (Livy 45.39.16). Single combat did, however, require the permission of the commander, so there were also limits to the scope for this particular form of heroic activity for young aristocrats.[42]

As for the rank-and-file, they could demonstrate courage in battle in a variety of ways. In the first place, there was the question of holding their ground and formation under enemy pressure, with the front line facing the horrors of close-order hand-to-hand fighting. A frequently cited illustration of this is the historian Sallust's description of the battlefield near Pistoria in northern Italy following the defeat of Catiline's uprising in 62 BC, which had been supported by many Roman veterans: 'Only when the battle was over could the courage and ferocity with which Catiline's troops had fought be fully appreciated. Virtually every man in death covered with his body the place he had held while fighting' (61.1–2).[43] But there was also scope for more pro-active, aggressive manifestations of courage in battle, whether it be the troops of Scipio storming New Carthage in 209, who 'propped their ladders against the wall and swarmed up them with undaunted courage' (Polyb. 10.13.10), or Caesar's men who, when engaging with Germans in 58, 'attacked with such vigour when the signal was given ... that there was no time to throw javelins at the enemy' (*B. Gall.* 1.52). Indeed, the behaviour of the two sides at the civil-war battle of Pharsalus in 48 illustrates both aspects, with Caesar's men advancing and delivering a volley of javelins before engaging, while Pompey's forces waited to receive the attack in the expectation that the advance would weary Caesar's men: 'They stood up to the hail of missiles, and bore the onset of the legions; they kept their ranks, threw their javelins, and then resorted to their swords' (Caes. *B. Civ.* 3.93).

Furthermore, certain aspects of Roman military practice during the Republic were designed to encourage ordinary soldiers to engage in individual acts of heroism, especially the award of *dona militaria* ('military gifts'). These took the form of symbolic decorations, such as a crown for the first man to scale the walls of a besieged city, and Polybius is explicit

[42] Oakley 1985 generally (permission at 407–8, McDonnell 2006: 204–5).
[43] Further discussion in Rosenstein 1990: 95–7, Phang 2008: 66.

that their purpose was to 'encourage young soldiers to face danger' and to 'excite [them] to emulation and rivalry in the field' (6.39.1, 8). Polybius' description of the practice refers only to a handful of such awards and indicates that those for wounding or killing an enemy were awarded only in specific circumstances, but it is apparent that during the late Republican period and beyond, the number of categories of award expanded, as did the circumstances in which one might qualify. It is also evident that these decorations could be won by ordinary soldiers.[44] Scipio's storming of New Carthage again provides an illustration: he is said to have 'promised crowns of gold to the first men to scale the walls, and the usual rewards to those who showed conspicuous bravery . . . [which, together with other factors] created great enthusiasm among the young soldiers and raised their spirits' (Polyb. 10.11.8). Polybius' comment about emulation and rivalry is corroborated by ancient accounts of battle which sometimes comment on soldiers excelling themselves in battle because they were aware that their commander happened to be well-placed to see them in action.[45] While such awards obviously served a practical motivating purpose, their symbolic character, and the fact that recipients were given precedence in religious processions and displayed their decorations in the reception area of their home (Polyb. 6.39.9–10), also reinforced the important place of courage in Roman cultural values. Another reflection of the high value placed on courage is the discussion in Roman sources as to who had displayed the greatest bravery in battle, with L. Siccius Dentatus, 'the Roman Achilles', from the fifth century BC, regularly a leading contender, on the strength of his having fought eight or nine single combats, bearing forty-five scars on the front of his body (but none on his back), and having been awarded many military decorations (which the sources enumerate at length).[46]

The kudos which accrued to individuals for acts of heroism in battle did carry potential risks for the overall success of Roman forces in an engagement. If too many individuals sought personal glory, this might jeopardise the coherence of the army's formations. The role of such discipline in

44 Maxfield 1981: 60–3, with Linderski 2001.

45 References from Caesar in Phang 2008: 47 n. 70.

46 Dion. of Hal. *Ant. Rom.* 10.36–7, Val. Max. 3.2.24, Plin. *HN* 7.101, Aul. Gell. 2.11.3. Although the historicity of Dentatus' achievements (and even of the man himself) is open to doubt (Oakley 1985: 393, Linderski 2001: 5–6), discussion of him in the sources still provides important evidence for Roman attitudes to courage.

Roman military culture may in fact have been overemphasised in modern discussions,[47] and in practice acts of individual heroism rarely seem to have precipitated disaster. This was perhaps due to the fact that the greatest scope for such bravery during the Republican period lay with those troops involved in skirmishing in front of the main lines (Polyb. 6.39.4) or with those in the cavalry.[48] The risks associated with such activities also need to be counter-balanced by their potential to inspire the rank-and-file to adopt an aggressive attitude when engaging with the enemy. It was a question of striking an appropriate balance between individualism and group identity, between initiative and obedience.[49]

There was also the potential for the manliness associated with military courage to be compromised by another aspect of military discipline – namely, the liability of soldiers to be punished by beating in the event of certain lapses, such as lagging behind on the march or losing equipment.[50] Beating was a punishment associated in Roman society above all with slaves,[51] so its use in a military context on freeborn citizen soldiers was potentially problematic – a point Tacitus exploits in his account of the military mutinies at the start of Tiberius' reign, where excessive corporal punishment fuelled soldiers' dissatisfaction (*Ann.* 1.16–44).[52] Although this apparent anomaly lacks a satisfactory explanation, the attractive suggestion has been made that soldiers' wounds on the front of their body, and the honour associated with them, were 'conceptually placed as the polar opposite of scars from servile beating' and functioned as a 'mark of manhood, the signifier, permanently inscribed on his body, of his social status as a full man.'[53]

The Principate witnessed some changes. Starting with its inclusion on Augustus' golden shield in the senate house (*RG* 34.2), *virtus* became associated particularly with the emperor (especially on coinage), and while it retained a core focus on military prowess, its association with service to the state was strengthened;[54] as the opportunities for aristocratic military achievement were reduced, Stoic thinkers began to redefine *virtus* in non-military terms;[55] the practice of formalised single combat disappeared with the end of the Republic;[56] and military decorations were increasingly

[47] Goldsworthy 1996, Phang 2008: ch. 2. [48] McCall 2002: 83–96.
[49] Discussions in Goldsworthy 1996: ch. 6, Lendon 2005: ch. 8. [50] *Dig.* 49.16.3.16, 49.16.14.1.
[51] Saller 1994: ch. 6. [52] Tac. *Ann.* 1.16–44, with discussion in Alston 1998: 216.
[53] Walters 1997: 40, with specific reference to Livy 2.23 where scars from wounds and from beating are juxtaposed on a soldier's body.
[54] Galinsky 1996: 80–84, Noreña 2011: 77–82. [55] Roller 2001: 97–108.
[56] The last recorded instance appears to be from 45 BC: Oakley 1985: 396.

awarded to officers rather than the rank-and-file.[57] At the same time, courage in military contexts continues to be important. *Virtus* appears as a quality worthy of commemoration in military epitaphs from the early centuries AD;[58] there are references to honourable wounds in a range of sources from the early Principate;[59] and there are episodes where individual soldiers displayed conspicuous courage, such as the trooper Longinus who, during skirmishing around Jerusalem in the course of the Jewish revolt of 66–70, dashed out of the Roman line into the middle of the Jewish forces, where he killed two of the enemy before returning to his own line (Joseph. *BJ* 5.312–16), or M. Valerius Maximianus whose epitaph recorded his having 'killed with his own hand Valao, chief of the Naristi', during the Marcommanic Wars of Marcus Aurelius.[60] Interestingly, however, these cases involved individuals serving in a cavalry unit, and in the case of Longinus, although 'his valour gave him distinction and led many to emulate his gallantry', his commander Titus subsequently tried to discourage such behaviour by 'ordering his troops to prove their manhood without running personal risks' (Joseph. *BJ* 5.316).

With the transition to Late Antiquity and emperors increasingly involved personally in warfare during the third and fourth centuries, the association of *virtus* with the emperor was intensified. The late third century rhetorical treatise on how to compose imperial panegyrics attributed to Menander Rhetor gives priority to the praise of military achievements, remarking that 'courage (*andreia*) reveals an emperor more than do other virtues',[61] and panegyrics from this period follow this advice. Pannonia, the region from which so many emperors in this period came, is praised for its reputation for *virtus* (*Pan. Lat.* 10.2.2), while the *fortitudo* and brave exploits of individual emperors are highlighted, as in the following comments on Constantine's early career (which, intriguingly, include a reference to single combat): 'by confronting the dangers of war and by engaging the enemy even in single combat you have made yourself more notable among the nations' (*Pan. Lat.* 6.3.3).[62] Inscriptions from the period also refer to imperial *virtus*, as in a dedication to Constantine and Licinius 'by whose courage and foresight barbarian peoples everywhere have been subjugated so as to the strengthen the protection of the frontier',

57 Maxfield 1981: 63–6, Phang 2008: 197–9.
58 Lendon 1997: 244 n. 42 for references.
59 Leigh 1995: 206.
60 *AE* 1956.124 (tr. Campbell 1994: 64–5), with discussion of these and other incidents in Goldsworthy 1996: 265–71.
61 Russell & Wilson 1981: 84–5 (s. 372, ll. 30–31).
62 See further Nixon & Rodgers 1994: 23 n. 85, Noreña 2011: 54 n. 62.

and another to Constantine 'outstanding in courage, victor and triumphator' (*ILS* 8938, 700).[63] The bravery of Constantius II in battle against the Persians is praised in a panegyric of the 350s (Julian *Or.* 1.25A), while Ammianus' obituary of Julian highlights his *fortitudo*:

> His courage is shown by the number of his battles and his conduct in war, as well as his endurance of extremes of cold and heat. In contrast to the physical efforts required of the common soldier, the work of a commander is mental, but Julian on one occasion boldly met and dispatched a savage foe, and often stemmed the retreat of our men single-handed by putting himself in their path. In his destructive campaigns against fierce Germanic rulers and on the burning sands of Persia, he would give his men confidence by fighting in the front rank. (25.4.10 [tr. W. Hamilton])

That knowledge of heroic ideals persisted into Late Antiquity is implied by Ammianus' reference to the individual achievements of the Republican heroes Marcellus and Siccius Dentatus (25.3.13), while the officers who led a band of soldiers on a daring night-time raid on the Persian camp during the siege of Amida in 359 were later honoured by Constantius II for their courage by the setting up of statues of them in full armour in Edessa (19.6.12). Meanwhile, when Julian wanted to punish troops who had deserted the ranks in the battle of Strasbourg in 357, he did so by symbolically questioning their manliness: he is said to have 'dressed them in women's clothing and led them through the camp to expel them, thinking this a punishment worse than death for many soldiers'. And apparently their sense of shame subsequently inspired them, when re-admitted to the ranks, to redeem themselves by fighting more bravely than others (Zos. 3.3.5). There is also some evidence to suggest a revival of the practice of formalised single combat in Late Antiquity,[64] but there is a risk of investing too much significance in it. The number of instances is far fewer than from the Republican period (across a comparable timespan); the Roman champions were very often individuals whose rather different cultural background may account for their enthusiasm for such exploits; and most of the episodes appear in the history of Procopius, who may have dramatised them 'partly for literary effect, partly because of his close connections with the very cavalry officers from whose ranks the champions emerged'.[65]

[63] Constantine was reportedly wounded in the thigh during his campaign against Licinius in 324 (*Orig. Const.* 24), suggesting there was some substance to this rhetoric.

[64] Lendon 2005: 302, 385 n. 32.

[65] Rance 2007: 377.

The etymological associations of *andreia* and *virtus* with masculinity prompt some final brief reflections on the involvement of prominent women in warfare. The historian Tacitus betrays elite male anxieties about female involvement in military affairs and in positions of authority in his comments on Agrippina the Elder's actions on the Rhine frontier in AD 15 when she is described as 'assuming the duties of a general' (*Ann.* 1.69), as also in his account of the senatorial debate in AD 21 about wives accompanying governors, which included critical reference to Plancina, wife of the governor of Syria, having 'presided at the drill of the cohorts and exercises of the legions' (*Ann.* 3.33). Against this background, it is unsurprising that the only prominent women who receive a measure of positive comment when acting in a military capacity are figures like Boudicca, Zenobia and Mavia – all exotic non-Romans from the peripheries of the empire.[66] Corroboration of this association can perhaps be found in the archaeological evidence for Roman face-mask helmets depicting female faces. Whatever the explanation for such helmets, they bear features associated with either Amazons or eastern cultures.[67]

[66] Boudicca (first-century British ruler): Tac. *Ann.* 14.34–7, Cass. Dio 62.1–8. Zenobia (third-century Palmyrene ruler): Zos. 1.39 ('had the courage of a man'), *HA, Duo Gall.* 13.3 ('braver and more skilful than many emperors'). Mavia (fourth-century Arab ruler): Sozom. *Hist. eccl.* 6.38 (led her forces in person and defeated a Roman general).

[67] Bartman 2005: 110–17.

CHAPTER 3

Manpower and Money

War-making has a close but complex relationship with resources. War can be a way for a state to acquire additional resources, but waging war also involves the consumption of the state's existing resources. In the Roman world, the most important of those resources was manpower. The first section of this chapter considers the mechanisms for recruitment over the course of Roman history, before turning to the demographic impact of military service. The second section shifts attention to another resource essential to war-making – money. As Cicero famously observed, 'the sinews of war – a limitless supply of money' (*Phil.* 5.5). The financial costs of the armed forces and of warfare over time are examined first, before the income which the Roman state gained from warfare is assessed.

3.1 Recruitment and Demography

Technological developments in weapons, armour and siege equipment influenced the conduct of warfare during Roman history (see further in Section 6.1), but there was never any technological innovation which revolutionised war in the way that the introduction of gunpowder or the invention of the aeroplane was to do in more recent periods of history. The outcome of warfare in the Roman world depended above all on the question of manpower – the ability of a state to marshal more men than the enemy. The effectiveness of the means by which the Roman state supplied manpower for its armies therefore assumed enormous importance.[1] The pressure to find enough men necessarily involved the use of compulsion, so that a history of Roman recruitment is at least partly a case of tracing the extent to which the state relied on conscription in different periods.

[1] Cf. Shaw 1999: 134.

The process of levying troops during the Republican period was referred to as the *dilectus*, whose literal meaning – 'choice' – might suggest a lack of obligation to undertake military service. However, the choosing was done by Roman officials from among those liable to serve, for whom the *dilectus* was a matter of compulsion.[2] In theory all citizens had an obligation to undertake military service, but as already seen (in Section 2.1), those with little or no property – the *proletarii* – were excluded from the legions, at least until the late second century BC. Any *assidui* – those with the requisite minimum of property – could theoretically be called up, but in practice those classified as *seniores* – those aged forty-six to sixty years – were not (save in an emergency), leaving the burden to be shared among the *iuniores* – those aged seventeen to forty-five. Those who fell in this group were in principle liable for sixteen years of service, but as previously noted, they could normally expect discharge after six years continuous service,[3] although this could be extended if circumstances demanded, such as during the Hannibalic War. In the early Republican period, the levy took place on the Capitol where officials would select the number of *iuniores* commensurate with the estimate of the forces required for that campaigning season and assign them to legions. With the passage of time, however, increasing numbers of citizens lived considerable distances from Rome, which must have forced the consuls to delegate much of the levying to regional Italian authorities. By the first century BC the levy of those living in or near Rome took place in the less restricted space of the Campus Martius, while recruiting officers (*conquisitores*) were sent to regions further from Rome.[4]

Confirmation that the process was underpinned by compulsion derives from a range of anecdotes concerning the penalties visited on individuals who tried to escape their obligation, from sale into slavery and exile to flogging and fines.[5] Also relevant to the issue of compulsion is the incidence of group resistance. Accounts of the early Republican period present plebeians as exerting pressure on the patrician elite for political concessions by withdrawing their military services en masse, but in the absence of even near-contemporary sources there is much uncertainty about the details.[6] During the better documented second century BC

[2] For discussion of the use of the term, see Brunt 1987: 635–8. [3] Brunt 1962: 80–81.

[4] Brunt 1987: 625–34, 726, Rawson 1971: 15–16.

[5] Brunt 1987: 391–2. Although the census took place only at approximately five-yearly intervals, it seems that the state kept track of boys under the age of seventeen to ensure they could be enlisted when they reached that age (Aul. Gell. 10.28.1).

[6] Cornell 1995: ch. 10.

there were a number of episodes involving resistance to the levy, arising from the unpopularity of a theatre of war, whether because of the perceived difficulty of the fighting or the limited prospects for booty; levies for Spanish campaigns particularly prompted opposition in this period. However, the fact that such episodes are reported implies that the levy did not usually meet with resistance and that *assidui* generally accepted their obligation of military service.[7] It was only in the final period of the Republic, when the Social War and then successive civil wars increased the demand for manpower to unprecedented levels that conscription took on a more consistently coercive form, as reflected in the activities of the *conquisitores* actively seeking out those eligible for service.[8]

In all this, 'soft' power also played an important role in Republican recruiting practices, especially the exercise of patronage. There is evidence that consuls could use their position to grant exemptions from service to those who would not otherwise have had grounds for seeking it, and that tribunes of the plebs could put pressure on the consuls to grant exemptions.[9] And commanders sometimes used ties of patronage to ensure the production of recruits, whether it be Scipio Aemilianus raising a force of 4,000 men for his Spanish campaign in 134 BC which included his clients and friends (App. *Hisp.* 84), or Pompey using his estates and family ties in Picenum to raise an army in the 80s (Plut. *Pomp.* 6).[10]

Augustus' formalisation of the shift from an army of part-time citizen conscripts to a full-time standing army had important implications for the conduct of recruitment, since it established the principle of redeploying existing troops to meet changing needs, rather than raising new legions.[11] Demand for manpower eased, compared with the pressures of the final decades of the Republican period, while the new option of the military as a form of permanent employment with a discharge bonus presumably encouraged volunteers. The fact that those volunteers increasingly did not come from Italy over the course of the first century implies that conscription was not being used, at least in that part of the empire.[12]

7 See Rich 1983: 316–18 for details and caution.

8 Brunt 1987: 408–10.

9 E.g., Livy 43.14 (169 BC), Livy *Per.* 48 (151 BC), 55 (138 BC).

10 Cf. Brunt 1990: 513, Whitby 1995: 66–8, both deploying instructive comparanda from eighteenth-century Scotland.

11 Rich 2003: 345–6.

12 Mann 1983: 49. When three legions were lost in the Varian disaster in AD 9 and not replaced, 'it was not that Italy could not supply the men, but the emperor shrank from the coercion that would have been necessary' (Dobson 1986: 13 n. 11); the decision is also likely to have been influenced by the opportunity it presented for major financial savings, especially with regard to discharge bonuses (Scheidel 1996: 135–7).

However, this does not mean that conscription was abandoned as one way of ensuring there were sufficient recruits in the provinces. All Roman citizens in principle remained liable for military service (*Dig.* 49.16.4.10), and there is a range of evidence to show that conscription was sometimes used during the Principate. However, that evidence mostly relates to the first century, while evidence emphasising the role of volunteers derives from the Severan period (late second and early third century). Such a pattern is consistent with military service becoming more attractive financially following Septimius Severus' increase in pay (by an unspecified amount) in the 190s, as well as a shift during the second century to more localised recruiting, which meant that soldiers tended to be stationed in the province from which they came.[13]

Conscription seems to have been a more consistent feature of the recruitment of provincial *auxilia*.[14] In AD 6, the requirement to contribute troops to Tiberius' German campaign was one factor which played a part in the drawn-out revolt of the Dalmatians (Cass. Dio 55.29), the Thracian revolt of AD 26 is reported to have been a reaction to Roman demands for military manpower (Tac. *Ann.* 4.46), while the revolt of the Batavi in AD 69 was likewise triggered by the mass conscription of their youth (Tac. *Hist.* 4.14). At the same time there is some evidence of Roman awareness of the benefits of harnessing the manpower of provincial societies by developing ties with local patrons, who could then be induced to place their military manpower at the disposal of the Roman authorities. This appears to have worked particularly, though not exclusively, in the western provinces.[15]

Another period of endemic civil war in the mid-third century, in addition to new military challenges on the eastern and northern frontiers of the empire, meant that demand for troops once again rose, with increased use of conscription a logical corollary, even if the meagre sources for this period offer little explicit confirmation. By the time more detailed evidence is available, from the emperor Diocletian's reign in the late third century, it is clear that conscription had again assumed major importance, not least because the empire's armed forces had expanded in size (see below). The main mechanism for conscription was the annual requirement that landowners provide recruits proportionate to the size of their estate;

[13] Brunt 1974. However, Rathbone 2009 argues that pay increases may have been doing no more than offsetting inflation, especially following the impact of the Antonine plague in the late second century.

[14] Saddington 1982: 140–1, Mattern 1999: 86, Haynes 2013: 99–100.

[15] Haynes 2013: ch. 7.

smaller landowners whose property did not meet the threshold for a single recruit were grouped together. These recruits came primarily from landowners' own tenants and reflect the way in which the recruiting system was based on the land tax assessments introduced under the Tetrarchy. In years when the levy did not require manpower from every province, the provision of recruits could be replaced in some regions by a tax known as the *aurum tironicum* (lit. 'recruit gold'), but this was a matter for determination by the government, not the landowners.[16] The system undoubtedly had its flaws, since, despite the existence of a vetting process for recruits, it proved possible for landowners to provide men whom they were keen to be rid of and so by implication were unlikely to make good soldiers (Veg. *Mil.* 1.7). Moreover, it seems that service was sufficiently unpopular for new recruits to be tattooed, and for physical violence sometimes to be required in the enlisting process.[17] Sufficient numbers resorted to self-mutilation in order to avoid service (most commonly, by cutting off a thumb) to warrant a string of imperial laws in the fourth century, although this was not an exclusively late antique phenomenon.[18]

Another symptom of the demand for men was the obligation of soldiers' sons to undertake military service when they reached adulthood, apparently also introduced by Diocletian.[19] Over the course of the Principate there had been an increasing tendency for the sons of soldiers to enlist, reflected in records by the designation *castris* ('from the camp'), but what had originally occurred on a voluntary basis now became a requirement. From the state's perspective, such an arrangement had the advantages of providing young men who had known nothing other than life growing up in a military context and who could be tracked through their youth: one law refers to children being inscribed on the register of their father's unit when still *infantes* (*Theod. Cod.* 7.1.14 [394]). It has sometimes been thought that with the contraction of the empire to its eastern half by the late fifth century, the pressures to employ conscription eased and that there was once again greater reliance on volunteering during the sixth century. However, this view depends heavily on the silence of the legal sources, which could equally be explained by their taking it for granted that fourth-century arrangements continued, which other scattered non-legal evidence, particularly from the later sixth century, suggests was the case.[20]

16 Jones 1964: 615. 17 Lee 2007: 80.
18 *Theod. Cod.* 7.13.4–5, 10; earlier case: Suet. *Aug.* 24. 19 Jones 1964: 615.
20 Whitby 1995: 75–87.

What numbers of men, then, did the Roman state have under arms in different phases of its history? This is a question relevant to a range of issues, above all the ability of the state to mobilise the manpower at its disposal, and the demographic impact of that mobilisation on the wider population. There is also the more specific question of the number of troops deployed on particular campaigns, which has implications for the logistical capabilities of the Roman state.

During the middle and late Republican periods the manpower on which the Roman state drew was mainly from the Italian peninsula (citizens and allies), although as Roman power expanded beyond Italy the contribution of *auxilia externa* in the final two centuries BC became more significant than is usually acknowledged (see Introduction, Section 3). However, it is above all the demands on the citizen body which can be assessed on the basis of the available data. Although larger forces could be mobilised when needed, as happened with a Gallic invasion of northern Italy in 225, the usual situation during the third century prior to the Hannibalic War was for four legions to be raised every year, two for each consul – so about 18,000 men in total.[21] During the Hannibalic War the number of legions levied rose dramatically from the usual four to about a dozen in the early years and then usually more than twenty from 214 onwards, with a peak of twenty-five in 212.[22] In the first half of the second century, the number of legions was lower, but still usually averaged ten, plus or minus two, per annum, now that Rome had regular commitments in Spain, northern Italy and the east, before dropping back into single digits most years in the second half of the second century, except when there were major threats such as those from the Cimbri and Teutones in the final decade.[23] The wars against Sertorius and Mithridates, Pompey's eastern campaigns and Caesar's Gallic war meant that the number of legions rose again into the twenties and sometimes higher during the middle decades of the first century,[24] while the civil wars at the end of the Republic saw the number balloon to the point where there may have been as many as sixty legions by the time of Actium.[25] These vastly increased numbers of legions in the first century BC are not, however, as significant as the bare figures might suggest, since the pool of available citizens had been increased very substantially by the grant of citizenship to the Italian allies following the

21 Rosenstein 2012a: 108. An additional 21,500 citizen troops were held in reserve at Rome in 225 (Polyb. 2.24).

22 Brunt 1987: 418 (Table X).

23 Brunt 1987: 424 (Table XI), 432–3 (Table XIII).

24 Brunt 1987: 449 (Table XIV).

25 Brunt 1987: ch. 26.

Social War of the 80s. So while the civil wars also placed a severe strain on citizen manpower, it was no worse than during the Hannibalic War and probably somewhat less severe when allowance is made for population growth over the intervening century and a half.

These figures for legions show that the Roman state during the Republic could mobilise much larger numbers of citizens than normal when circumstances required, but there remains the question of the larger demographic significance of these figures – that is, what proportion of the citizen population did they represent? Answering that question is not straightforward, firstly because ancient sources give varying sizes for the legion across this period of Roman history, with a range of figures between 4,000 and 6,000 men,[26] and secondly because the ancient evidence allows different estimates of the citizen population of Republican Italy. The latter is contentious because of disagreement about how to interpret the primary evidence for the subject, namely the surviving figures from the census of citizens normally taken every five years.[27] Their most intensely debated feature is the dramatic increase from less than one million in 70/69 BC to more than four million in 28 BC. Some scholars, such as Peter Brunt, have accounted for this by assuming that the former (and earlier) figures represent only adult males, while the latter must also have taken into account women and children. Once an estimate for women and children is removed from the latter figure, this would leave, it is argued, a figure of about 1.7 million adult male citizens in 28 BC.[28] This has been dubbed the 'low' count, in contrast to the view of others, such as Elio Lo Cascio, that the census figure of four million for 28 BC was, like earlier census figures, only adult males, thereby giving a 'high' count. The issues and arguments are too complex for elaboration here, but a majority of scholars have inclined towards the 'low' count, although a 'middle' count has also been proposed, so it is not a binary choice.[29]

The 'low' count approach gives a figure of 29 per cent of male citizens serving in the legions at the time of peak demand during the Hannibalic War in 212, which is a very high mobilisation rate, but not impossible. When it is converted into a percentage of total citizen population (including women and children), it yields of figure of 12.6 per cent, which is comparable to the mobilisation rate of about 11 per cent by the Confederate States during the American Civil War in the 1860s. Even using a

[26] Discussion in Brunt 1987: 687–93, Roth 1994.
[27] Brunt 1987: 13–14 tabulates the figures and sources. [28] Brunt 1987: 117.
[29] Summary of the debate in Scheidel 2008: 19–30; Hin 2013 proposes a 'middle' count.

'high' count approach, this yields a figure of 6.4 per cent of the total citizen population in 212, which is comparable to mobilisation rates in eighteen-century Prussia and Sweden. Either way, Roman mobilisation rates during the Hannibalic War were demographically feasible, while also being higher than those achieved by most western European states in the eighteenth and nineteenth centuries, regardless of which interpretation of the Roman census data is correct.[30] While Rome would no doubt have struggled to maintain such rates over more than a few years, it appears to have had the capacity to achieve very high mobilisation rates in crisis situations, such as the Hannibalic War or the Social War.

During the Principate the parameters relating to manpower differed significantly from the Republic. First, the contribution of the population of Italy to the legions declined whereas provincial populations, which constituted a numerically much more substantial pool of manpower, were exploited more consistently together with some use of manpower from beyond the frontiers. Secondly, Augustus' formalisation of a standing army implied a more consistent number of men under arms, albeit with occasional variations arising either from the loss of legions or the creation of new legions to strengthen particular sectors of the empire. So the twenty-five legions which Augustus left at his death in AD 14 rose to thirty-three by the time of Septimius Severus, nearly two centuries later, during which time new provinces in Britain, Arabia, Dacia and Mesopotamia had been added to the empire, creating the need for more troops (while also providing additional manpower). Calculations of the total size of the armed forces during the Principate vary from approximately 400,000 to 500,000 men. These variations depend partly on the point in time chosen, since (as noted) the number of legions increased, as also did the number of *auxilia*, and partly also on the view taken (again) about the size of the legion (ranging between 5,000 and 5,500 men). However, the most important factor in explaining the variations is assumptions about the proportion of *auxilia* to legions, where there is greater scope for uncertainty, with a majority of overall estimates gravitating towards *c.*400,000.[31]

Where does this stand as a proportion of the empire's population during the Principate? The evidence for population size is meagre, with estimates having to rely on a range of proxy data, above all the likely carrying

[30] Figure of 29 per cent: Hopkins 1978: 33, based on Brunt's data; other percentages and comparisons with more modern states: Scheidel 2008: 38–41, 65.

[31] By way of example, see Campbell 2002: 176 (*c.*378,000, end of 2nd c.), Potter 2014: 126 (*c.*400,000, end of 2nd c.), Birley 1983: 43 (*c.*430,000, mid–2nd c.), Hassall 2000: 321 (*c.*500,000, early 2nd c.).

capacity of regions and comparative evidence of population densities in more recent periods of history (which are assumed to provide upper limits for the Roman period), but it is broadly accepted that the empire's population was approaching 50 million inhabitants by the time of Augustus, rising to perhaps 60 million by the middle of the second century, before the impact of the Antonine plague.[32] Irrespective of whether the empire's military forces comprised 400,000 men or 500,000, it will have accounted for no more than one percent of the empire's population. Of course, since women were not eligible for military service, the available pool of manpower needs to be halved, raising the proportion to two per cent, while discounting children and the elderly would push that figure higher. How much higher depends on assumptions about mortality rates in the Roman world, about which it is difficult to be specific. Overall, however, an army of half a million men should not have been too heavy a demographic burden on an empire of 50 million inhabitants, and stands in stark contrast to the much higher military participation rates achieved at various points during the Republic (see above).

As for Late Antiquity, it is generally accepted that the armed forces increased in size in the fourth century, but estimates still range from *c.*500,000 to *c.*600,00.[33] If one accepts an estimate of *c.*400,000 for the second century, and one of *c.*500,000 for the fourth century, then this would represent an expansion of about 25 per cent, which one might have thought would impose a heavier burden on the empire's manpower, especially if one also factors in some degree of demographic contraction arising from the Antonine plague in the later second century and also the impact of the so-called Cyprianic plague of the mid-third century.[34] However, even allowing for these – and a resumption of population growth in the fourth century should not be discounted – estimates of mobilisation rates in Late Antiquity do not suggest that recruitment exerted excessive pressure on the available manpower.[35] Even the major pandemic which swept through the empire in the mid-sixth century does not seem to have had a significant impact on recruitment, perhaps because its worst effects were felt in urban centres, whereas the military drew

[32] Frier 2000: 811–16, Scheidel 2001b: 63–70, 2007: 47–9.

[33] Overview of data and arguments in Lee 2007: 74–77; for scepticism about any significant increase, see Harris 2016: 229–31.

[34] Duncan-Jones 1996, Harper 2015; for scepticism about the impact of these plagues, see Haldon *et al.* 2018 (Part 2).

[35] Carrié (1986: 470–8) estimates that between 1 in 19 and 1 in 23 men between twenty and forty-five years of age was serving in the fourth-century army.

primarily on rural populations.[36] The main manpower issue remained that when significant numbers of soldiers were lost in particular battles or campaigns, as happened more often in Late Antiquity, it was difficult to replace trained and experienced troops quickly.[37]

Alongside the broad question of army size in relation to available manpower, there is also the issue of the size of campaign armies in different periods of Roman history. This is of interest as an index of the logistical capabilities of the Roman state. As with determining the overall size of the Roman state's armed forces, there are the same methodological problems when it comes to figures for individual campaigns. Although larger armies were sometimes assembled, such as the 70,000–80,000 troops who faced Hannibal at Cannae,[38] armies of 25,000–30,000 men feature more regularly in a range of contexts from the middle Republican period.[39] In the late Republican period, Crassus invaded Parthia in 53 BC with an army of about 45,000 men (Plut. *Crass.* 20), while Caesar deployed a similar number at the siege of Alesia in Gaul in 52 BC.[40] The largest concentrations of troops during the Principate included the army assembled for the campaign against Maroboduus in AD 6, which was something of the order of 100,000 men, as also the forces with which Trajan invaded Dacia on his second campaign in 105.[41] From Late Antiquity, the largest credible force from the fourth century was Julian's army of 65,000 with which he invaded Persia in 363 (Zos. 3.13.1), while the largest known army from the sixth century was that of 52,000 sent by Anastasius in 503 in response to the Persian invasion of northern Mesopotamia (Josh. Styl. 54). Armies of up to 50,000 men are consistent with the size of later pre-industrial European armies, before the Napoleonic period saw a significant increase in size, with logistical limitations the main constraint.[42] The fact that the Roman state was sometimes able to support larger campaign armies is a reflection of the relative degree of organisational support which it was able to develop over time, and perhaps also the fact that it relied predominantly on infantry rather than cavalry, whose maintenance posed greater logistical challenges. At the same time, the relative rarity with which forces in excess of 50,000 men feature shows that that organisational support had definite limits.[43]

36 Whitby 1995: 92–103. 37 Lee 2007: 77–9. 38 Lazenby 1978: 79–80, Daly 2002: 25–9.

39 Scipio's force to Africa (204): Keppie 1984: 31. Cynoscephalae (197): Livy 33.3.9–10. Magnesia (190): Livy 37.39.

40 Keppie 1984: 92. 41 Dobson 1986: 18. 42 Nicasie 1998: 203–7.

43 For Roman military logistics in different periods, see Goldsworthy 1996: 287–96, Erdkamp 1998, Roth 1999, Lee 2007: 85–98, McCunn 2018.

3.2 Financial Costs and Benefits of War

The original principle on which Roman military service was based was that citizens bore the cost themselves. There was no pay for military service in the early centuries of Roman history when campaigning was localised, and citizens were expected to provide their own armour and weapons according to their financial ability, as determined by the regular census. However, as Rome's campaigns took armies further afield within the Italian peninsula for longer periods, a daily allowance (*stipendium*) was introduced to cover living costs. Ancient sources give the date of its introduction as 406 BC during the war against the Etruscan city of Veii, but doubts about their reliability for this period of Rome's history and the fact that Rome did not yet have a system of coinage leave scope for doubt about this date; however, if 406 was not the date, then the Samnite wars of the fourth century are a plausible context. More importantly, the sources link the introduction of the *stipendium* with the first levying of a property tax, the *tributum*, with the implication that the latter funded the former.[44]

Although there is debate about precise figures, one plausible calculation has estimated the total cost of the *stipendium* at 228 million *denarii* during the first half of the second century BC, rising to 968 million *denarii* during the first half of the first century BC.[45] This dramatic increase is accounted for partly by the much larger number of legions under arms during the Social War and civil wars of the 80s and 70s and partly by the state having to assume full responsibility for paying allied units after the Social War. Taking into account all the state's likely military costs (i.e., including matériel and navy), they represented about 60 per cent of the state's income during the first half of the second century BC and 65 per cent during the first half of the first century BC.[46] The modest increase in the latter percentage, despite the tripling of the outlay on the *stipendium*, is consistent with military costs increasingly being offset by the material benefits of successful warfare during the second century and beyond. The first indication of this came in 167 when the acquisition of the treasury of the kings of Macedon after the battle of Pydna allowed the suspension of the *tributum*. That the *tributum* was not reintroduced later in the second century (or thereafter – except during civil wars) is presumed to reflect the state's ability to fund the *stipendium* from the regular and

[44] Eich & Eich 2005: 20–1, Rich 2007a: 18, Northwood 2008: 265 n. 30, Rosenstein 2012a: 107.
[45] Kay 2014: 25–6, 299–300.
[46] Based on the figures for military costs and state income given in Kay 2014: 301.

increasing income derived from the taxation of newly acquired provinces (see further below) – and, it has been argued, the concern of the senatorial elite no longer to be beholden to ordinary citizens and their *tributum* when it came to making decisions about war and peace.[47]

Augustus' formalisation of a standing army created a substantial continuing financial burden for the Roman state. That burden arose primarily from the fact that Augustus' arrangements included provision of a substantial discharge bonus for soldiers who completed their term of service – 12,000 *sestertii* (HS) for legionaries and 20,000 HS for praetorians. Early in his reign discharge bonuses had taken the form of land grants with cash donatives, but because of the difficulties of finding sufficient land in Italy without resorting to politically unacceptable confiscations, it was changed to a money bonus from 13 BC, initially funded out of his own pocket. In AD 6, no doubt realising that even his vast wealth was inadequate to sustain such an expense in the long term, Augustus shifted the burden to the state by establishing a new military treasury (*aerarium militare*).[48] Although he contributed a significant cash injection at the outset,[49] the treasury was primarily funded through new taxes – a 5 per cent levy on inheritances and a 1 per cent levy on sales by auction in Italy, which will mostly have applied to sales of property – 'in effect charging landowners in Italy instead of confiscating their land'.[50] These new taxes were not popular, especially with the elite, but the institution nonetheless functioned until the early third century.[51]

Determining the scale of the overall burden arising from the formalisation of a standing army is beset by difficulties. Some of the relevant figures for army costs are known, such as legionary pay until the end of the first century (900 HS *p.a.* from Augustus' day until Domitian increased that amount by a third to 1200 HS in 84), or can be determined within relatively small margins of variation, such as the size of a legion (between 5,000 and 5,500).[52] However, there are also many other factors which are much less certain, such as the number of auxiliary units and their pay, how far actual unit strengths fell below paper strengths, the pay of officers, and the proportion of soldiers who survived their term of service to collect their discharge bonus. As a result, there is some variation in estimates of the

47 Eich & Eich 2005: 29, Tan 2017 (for the latter point). Revival in civil wars: Brunt 1987: 21.

48 It has been suggested that 'the annual cost of discharge was equivalent to the cost of paying the *stipendia* of almost ten extra legions (eight after the extension of service to twenty-five years)' (Dobson 1986: 14).

49 *RG* 17.2 (170 million HS). 50 Rathbone 2007: 175. 51 Corbier 1974.

52 Le Bohec 1994: 24 (5,000), Duncan-Jones 1994: 34 (5,500), Campbell 2002: 175 (5,240).

overall annual military budget in the first century, ranging from 370 million HS, to 450±50 million HS, to 493 million HS.[53] Furthermore, these figures need to be contextualised within the overall size of the budget which Roman emperors had at their disposal, which is even more uncertain.[54] Estimates of the budget range from 670 million HS to 900 million HS.[55] Depending on which set of figures one chooses, military expenditure could have accounted for anything between 25 per cent and 75 per cent of the total budget, but various considerations point towards the lower half of that range. First, the primary attraction of military service as a form of employment was its guaranteeing a regular income, but that income itself, while adequate, was not generous, especially when deductions for food and equipment are taken into account[56] – and since conscription remained an option for the government, it did not need to be. Moreover, the fiscal regime which prevailed during the Principate was one of low taxes,[57] with estimated rates of less than 10 per cent, so the state could not afford to be consistently liberal towards troops. Secondly, while the military was likely to have been the largest budget item, emperors did have other significant financial outlays to consider, above all expenditure on bread and circuses for Rome and on public building works more widely.[58] Since the emperor Septimius Severus increased military pay by an unspecified amount, and then his son Caracalla raised that again by another half,[59] and since the number of legions and auxiliary units had increased by the end of the second century, one might have expected the burden of the military budget to have increased. However, these additional costs are likely to have been offset by growing tax income arising from the addition of new provinces over the course of the second century (Arabia, Dacia, Mesopotamia) and by some degree of economic growth during the Principate.[60]

It was only with the increased incidence of warfare on multiple fronts during the mid-third century, and the disruption to revenue streams arising from the loss of provincial territory, whether permanent (Dacia)

[53] Duncan-Jones 1994: 36, Campbell 2002: 84, 175–6, Hopkins 2002: 199 n. 15. Cf. also Rathbone 2007: 173–4 (where the figures in *denarii* should be multiplied by 4 to give equivalents in *sestertii*).

[54] 'The only known element in the Roman state budget is the total cost of the army; perhaps 'known' is an exaggeration, but there is in broad terms a rough agreement among modern scholars that army cost say 450–500 million HS in the middle of the first century' (Hopkins 2002: 199).

[55] Duncan-Jones 1994: 33–46, Hopkins 2002: 200.

[56] Rathbone 2007: 163–5, 2009: 310–12.

[57] Hopkins 2002: 201.

[58] Rathbone 2007: 175.

[59] Campbell 2002: 84–5, Rathbone 2007: 159–60.

[60] Rathbone 2007: 175–6. For reasons to think there was modest economic growth during the Principate, see Saller 2002, Lo Cascio 2007: 619–22.

or temporary, together with inflationary pressures, that the mechanisms for funding the military underwent major changes. An emergency shift to taxation in kind during the mid-third century was formalised by the emperor Diocletian into a new fiscal system whose designation as the *annona* ('grain') is a clear indication of its fundamental principle. Although this system of taxation in kind gradually reverted during the fifth century to one involving taxes paid in money, the disruptions of the third century, and the adoption of a new coinage system in the fourth, based on the gold *solidus*, make it difficult to compare Late Antiquity with the Principate with respect to the military budget. The fourth-century army was larger, though again there is debate about the extent of the increase, with various options ranging from 500,000 to 600,000 (see Section 3.1), but this might have been offset by lower pay. In the more politically unstable conditions of Late Antiquity, the granting of donatives – money gifts to mark an emperor's accession and the five yearly anniversaries of accessions – became more regular than had been the case during the Principate, adding to government costs. On the other hand, discharge bonuses probably became less important. The disappearance of the military treasury in the early third century may simply have been a consequence of wider administrative reorganisation arising from the reforms of Diocletian, but it could also point to monetary discharge bonuses declining in significance. Consistent with this is increasing reference in fourth-century sources to discharge bonuses taking the form of land. However, the benefit to veterans which receives the most emphasis in the sources is exemption from, or reduction of, various taxes,[61] which while not cost neutral, did not require actual expenditure. Needless to say, just as the patchy evidence from the Principate allows plenty of scope for variations in estimates of the military budget as a proportion of the total budget, so also in Late Antiquity, there is a similarly broad range of opinion, not aided by the complication for estimates which the loss of the western half of the empire in the fifth century creates. Once again, one can find estimates of two-thirds or more at one end of the spectrum, with a third at the other end, and many opting for about a half.[62] Although complaints in the sources about the heavy burden of taxation in Late Antiquity probably involve an element of exaggeration, it is likely that there was a move away from the low tax regime of the Principate, which must reflect in part the increased military challenges that the empire faced during Late Antiquity.[63]

[61] Jones 1964: 635–6. [62] Treadgold 1995: 172, Hendy 1985: 172, Wickham 2005: 73–4.
[63] Wickham 2005: 62–72.

Turning in more detail to the financial benefits of war, ancient states benefited financially from successful war-making in two main forms. First, there were short-term gains from the acquisition of mobile forms of wealth, while secondly, there was the acquisition of land which could then be exploited in various ways in the longer-term and on a more lasting basis. The remainder of this section aims to map out continuities and changes in these two aspects across the main periods of Roman history.

The most obvious form of short-term gain was booty, about which more will be said shortly. But this heading also includes so-called indemnities whereby, particularly during the middle Republican period, the Roman state required defeated enemies to make payments towards the cost of the relevant war. The largest recorded sums were the 10,000 talents of silver from Carthage after the Hannibalic War and the 15,000 talents of silver from Antiochus III in 188 BC (although Carthage's payments were strung out over fifty years as a form of control, whereas Antiochus' were over a much shorter period). During the third century, Rome had acquired at least 4,675 talents of silver from defeated enemies, but the indemnities from Carthage and Antiochus helped to boost indemnity income during the first half of the second century BC to at least 27,280 talents (one consequence of which was that the censors of this period were able to fund unprecedentedly ambitious building programmes in the city of Rome). Although imposition of indemnities became less frequent thereafter, in the mid-80s BC Sulla required Mithridates to pay 2,000 talents as part of his settlement with him, and the cities of the province of Asia which had received Mithridates' troops to pay the huge amount of 20,000 talents. Just how substantial these sums were can best be appreciated by considering some comparanda: maintaining one legion for a year in the second century BC cost approximately 100 talents, while the most expensive known building project undertaken during the Republican period – construction of the 91 km long Marcian aqueduct in the 140s BC – cost 7,500 talents.[64]

Indemnities were a phenomenon of the middle Republican period, but booty remained a constant feature of Roman warfare, although the scale of opportunities varied and data for the value of booty are more problematic than for indemnities. Accounts of some triumphs during the Republic include figures for the precious metals paraded through the streets of Rome

[64] Kay 2014: 24 (comparanda), 37–42 (tabulation and discussion of recorded indemnity figures for third and second centuries). Mithridates' and Asian indemnities: Plut. *Sull.* 22, 25. Building projects: Rosenstein 2012a: 246–7.

as part of the booty, totalling amounts in excess of 500 talents, while numbers for war captives sold into slavery can yield estimates based on typical slave prices. Even at a conservative price estimate, the notorious enslavement of 150,000 Epirotes in 167 BC would have produced at least 1,700 talents.[65] As for non-human booty, Roman plundering 'brought immobilized and sequestered wealth, such as state reserves and temple treasures, into circulation again and put them to productive use'.[66] Of course not all of this went into the state treasury, since generals were expected to reward their troops and were also entitled to claim some of the booty for themselves, although this may have been no more than about 5 per cent of the total.[67] Even so, a substantial amount must have ended up in the treasury, and while much of this enhanced income was then spent on funding the legions to fight further wars, it also had a multiplier impact on the development of the Roman economy during the later Republican period, reflected in evidence for growth in the volume of output and exchange.[68]

Opportunities for major hauls of material booty decreased during the Principate, as the empire mostly faced enemies less wealthy than Carthage and the Hellenistic kingdoms of the Republican period, though there were exceptions: the suppression of the Jewish revolt in 70 appears to have provided sufficient booty to fund the construction of the Colosseum in Rome, while Trajan's forum complex was built from the proceeds of Dacian booty.[69] However, the relative lack of economic development among many of the empire's enemies during the Principate was less limiting on the quantity of available human booty. A total figure of 97,000 prisoners is given by one contemporary for the Jewish War, while a wealth of other evidence survives for slaves from that and other conflicts during the Principate, implying that war remained a more important source of slaves during this period of Roman history than has usually been acknowledged.[70] Likewise in Late Antiquity, it was still sometimes possible for substantial numbers of war captives to be enslaved by Roman armies,[71] and for substantial quantities of material booty to be acquired, especially when Romans forces managed to capture the baggage train of the Persian king (as they did on a number of occasions) and when Belisarius took possession of the Vandal and Gothic treasuries in 534 and 540.[72]

[65] Mattern 1999: 152–3, Kay 2014: 30. [66] Kay 2014: 279–80. [67] Kay 2014: 30–7.
[68] Kay 2014: ch. 11; cf. Hin 2013: ch. 2.
[69] Mattern 1999: 150–2. Colosseum: Alföldy 1995 with Millar 2005. Trajan's forum: Aulus Gellius 13.25.1.
[70] Bradley 2004. [71] Lenski 2011a: 193. [72] Lee 2007: 102–3.

However, these gains need to be offset against the greater frequency of inhabitants of the empire being carried off into slavery by Persian and barbarian invaders and raiders, along with material booty.[73] Moreover, in the fifth and sixth centuries it became common for the Roman authorities to give money, either as one-off or regular payments, to the Persians, Goths, Huns and Avars, in order to maintain peace. These 'subsidies' or diplomatic 'gifts', which attracted considerable criticism from contemporaries because of their potential connotations of subservience, could be viewed as the inverse of the indemnities extracted from Carthage and Antiochus in the Republican period. Although they sometimes involved quite substantial amounts, they could still be considered an economical alternative to the costs of waging war and of defeat.[74]

As for the acquisition of resources on a more permanent basis, the middle Republican period saw the Roman state gain control of significant areas of land within the Italian peninsula through confiscation of the land of defeated enemies, chiefly during the conquest of Italy (338–265) and from former Italian allies who deserted the Roman cause during the Hannibalic War. Some of this land was effectively privatised through the establishment of colonies or grants to individual citizens – one of the main ways in which ordinary Romans benefited from successful war – and would have generated some tax revenue. However, the amount of land confiscated is difficult to determine, and substantial areas remained 'public land' (*ager publicus*), which Italian allies were allowed to continue farming in return for rent or pasture dues – an arrangement which helped to strengthen their loyalty to Rome, at least until the late second century BC.[75]

The middle (and late) Republican period also witnessed the steady acquisition of increasingly significant territory and property outside the Italian peninsula – on the one hand, provinces, and on the other, specific property of economic value such as mines. Particularly important from the point of view of immediate income was the gaining of Iberian silver mines in the early second century BC which a range of evidence, including levels of atmospheric lead pollution from silver extraction preserved in Greenland ice core samples, suggests were quickly exploited on a large scale. By the middle of the second century BC, the Spanish mines were reported to be producing 1,500 talents of silver per annum, supplemented by income

[73] Lee 2007: ch. 4, Lenski 2011a: 186–92.
[74] Lee 2007: 119–22. For subsidies during the Principate, see Mattern 1999: 121, 159–60, 180.
[75] Roselaar 2010.

from Macedonian gold and silver mines.[76] As for the economic benefits of provinces, charting tax income during the Republic is problematic because of uncertainties about figures and the speed with which the state exploited opportunities in different regions. The Romans began levying taxes on Sicily and Sardinia by the end of the third century, followed by Spain in the early second,[77] but it was the acquisition of the wealthy province of Asia (western Anatolia) which appears to have prompted a more systematic approach to taxation from the late 120s.[78] By the time of Pompey's eastern conquests in the 60s BC revenues may have reached as much as 135 million *denarii* or 22,500 talents per annum, allowing (among other things) the abolition of harbour dues throughout the Italian peninsula[79] – though much of this wealth ended up in hands of the elite, rather than the state's coffers.[80] The acquisition of Egypt in 30 BC was a significant new addition to the empire's economic resources because of the extraordinary agricultural productivity of that region, but thereafter territorial expansion became increasingly more limited, with Dacia and its gold mines being perhaps the only further gain of major economic significance (early second century).[81] At the same time, administrative oversight of mines appears to have been streamlined during the Principate in order to maximise revenues, even if the bulk of that revenue came from leasing mines to private companies.[82]

The limited permanent loss of territory, and hence of tax revenues, during the third century – only Dacia and the Black Forest region of Germany – helps to explain why the empire was able to recover strongly in the fourth century, and the empire's retention of Egypt until the early seventh century is important in explaining the longevity of imperial power in the east. By contrast, the loss of economically valuable north Africa to the Vandals in the 430s, and the failure to recover it despite repeated efforts during the mid-fifth century, goes far towards accounting for the growing inability of western emperors to sustain political and military authority over the western provinces. The longer the western treasury had to manage without revenues from north Africa, the greater the difficulty of maintaining armed forces on a scale capable of rectifying the situation. Although Justinian's reconquest of north Africa and Italy in the 530s and 550s respectively had the potential to strengthen the state's economic resources significantly, scholars have tended to be cautious about

76 Kay 2014: ch. 3. 77 Kay 2014: 75–6. 78 Kay 2014: ch. 4.
79 Plut. *Pomp.* 45 with Harris 2007: 514. 80 Tan 2017: chs. 2–3. 81 Mattern 1999: 154–6.
82 Hirt 2010, Mattingly 2011: ch. 7.

the fiscal impact of these territorial gains, not least because of the economic damage arising from the length of the war in Italy and ongoing turmoil in north Africa.[83]

From the perspective of the Roman state, then, the economic impact of warfare varied from period to period, depending on location, duration and outcome of conflict, and on the economic development and potential of opponents. During the Republic, the Hannibalic War and the civil wars of the late Republican period were lengthy and economically damaging to Italy, but their negative impact was more than offset by successful overseas wars, especially against the wealthy Hellenistic kingdoms (which, surprisingly, succumbed relatively quickly and so with minimal damage to their economic infrastructure), but also less developed regions like the Iberian peninsula with significant mineral resources. The economic resources which flowed to Italy as a result had an economic impact many times greater than their individual parts, showing that despite its inherently destructive nature, war could facilitate significant economic growth.[84] Opportunities for comparable gains were limited during the Principate, and while the demise of the west in the fifth century provided Justinian with the opportunity for substantial conquests in north Africa and Italy in the sixth, any economic benefits were counterbalanced by the destructive duration of the Italian war and persistently unsettled conditions in north Africa after its initially rapid conquest. Needless to say, the contraction of imperial control over the west during the middle quarters of the fifth century represented a significant loss of economic resources by the empire, as even more so the conquest of Egypt and Syria by the Arabs in the early seventh century. These developments leave no doubt as to war's capacity also to have a serious negative economic impact.

[83] Lee 2007: 104–5. [84] Bang 2013, Kay 2014.

CHAPTER 4

Authority and Allegiances

Armies are necessarily authoritarian institutions in which a high premium is placed on obedience, discipline and order, and the extent to which these values can be instilled has a significant influence on effective performance in war. This chapter examines some of these issues in the context of Roman warfare. The first section considers the role of leadership, which can play a crucial part in motivating troops to obey willingly, by examining Roman ideals with regard to generals, and ways in which generals maintained their authority. The second section investigates strategies by which the Roman state sought to ensure obedience on the part of soldiers, alongside causes of unrest and how it was dealt with. The third section reflects on the phenomenon of civil war in the Roman world and how soldiers were induced to fight against other soldiers who also claimed allegiance to the Roman state.

4.1 Leadership and Command

What did Romans regard as the ideal qualities of a general?[1] One succinct statement was provided by the late Republican orator Cicero when advocating the claims of Pompey to be assigned command of the war against Mithridates of Pontus in 66 BC: 'In my opinion the greatest general needs to have four qualities – knowledge of military matters, courage, a reputation for leadership and good fortune (*scientia rei militaris, virtus, auctoritas, felicitas*).'[2] While this statement, and its subsequent elaboration, occurs in a tendentious context, it nonetheless provides a useful starting point for

[1] For reasons of space, the focus here is on generals, but this is not to deny the importance of officers and centurions as leaders.

[2] *Leg. Man.*28. Cf. (from the fourth century) *Pan. Lat.* 2.6.1 (*virtus* and *felicitas* as the two requirements of outstanding commanders).

examining Roman ideals in this area.[3] Although 'knowledge of military matters' might potentially imply an educational aspect, there was no formal training programme for generals or officers at any point in Roman history,[4] and indeed Cicero's elaboration of this quality with respect to Pompey proceeds to focus on his education through practical experience (in a career, it should be added, which can hardly be considered typical). Nonetheless, the inclusion of this quality in Cicero's list provides an indication that military competence, however acquired, was regarded as important. Similarly, in the fourth century the emperor Julian was praised by the historian Ammianus Marcellinus for his knowledge of military affairs (*castrensium negotiorum scientia*), defined in terms of various practical skills – effective conduct of sieges and deployment of troops in battle, careful selection of sites for camps, wise placement of advance posts (25.4.11). Commanders with limited military experience could offset this by ensuring they were accompanied by experienced deputies, as Cicero did during his period as governor of Cilicia in the late 50s BC.[5]

Cicero's second quality – courage – has already received attention (Section 2.2), although Cicero's elaboration of *virtus* with respect to Pompey shows that he interpreted it more broadly as incorporating integrity in a general's dealings with his own men (e.g., not selling commissions) and with non-combatant communities (e.g., restraining troops from plundering civilians with whom they are quartered) (37–8). Nonetheless, it is apparent that the willingness of a general to demonstrate personal courage could be important in reinforcing his authority among the rank-and-file and in instilling morale. During a setback in the fighting for control of Rome in 88 BC, Sulla is said to have done just this by seizing a standard and risking his life in the front line (App. *B. Civ.* 1.58), as did Caesar when one of his legions got into difficulties during a battle with the Nervii in 57 BC. Grabbing a shield, he moved through the ranks towards the front line, shouting instructions and encouragement. 'His coming gave them fresh hope; each man wanted to do his best under the eyes of his commander-in-chief, however desperate the peril' (*B. Gall.* 2.25). Yet while personal involvement by a general in fighting could help to turn the tide, it was of course potentially hazardous, with wounding or death of

[3] Onasander's treatise on generalship might have seemed a more obvious starting point, but its generality and philosophical focus limit its value for this subject: see Smith 1998: 158–65, Campbell 2004: 65.

[4] Campbell 1987: 22. [5] Goldsworthy 1996: 132.

a commander invariably the signal for panic and flight.[6] Unsurprisingly, therefore, there was a strand of thinking which counselled against generals risking their personal safety unnecessarily. Whatever its limitations, the first-century military handbook of Onasander advised that the general should 'keep entirely out of a hand-to-hand fight with the enemy – for even if in battle he shows unsurpassable courage, by fighting he cannot bring a benefit to the army comparable to the loss if he is killed' (33). In the early fourth century, a panegyrist ventured to voice muted criticism of Constantine, 'upon whose life the fates of all men depend', for endangering his life in battle (*Pan. Lat.* 12.9.4–6); and the late antique historian Theophylact Simocatta reported that when the general Philippicus tried to position himself in the forward ranks at the battle of Solachon (586), his officers persuaded him to withdraw to a safer position (2.3.10–12). Skilled generals sometimes found ways to demonstrate their courage while minimising their exposure to danger, as when, during the siege of New Carthage, Scipio was protected from enemy missiles by three shield-bearers while he inspected the forward lines – 'but the fact that he was in full view of his men inspired them to fight with redoubled spirit' (Polyb. 10.13.5). In a similar vein, Caesar had all the officers' horses, including his own, sent away prior to a battle 'so that everyone might stand in equal danger' (*B. Gall.* 1.25.1) – or at least appear to, since Caesar will presumably still have been positioned to the rear.[7]

The third quality identified by Cicero – reputation for leadership (*auctoritas*) – meant, in Cicero's presentation of Pompey's case, the impressive record of success that he had already acquired by the relatively young age of forty, about which it is therefore less easy to generalise. However, *auctoritas* was not a quality restricted to the military field and could be demonstrated in other contexts, above all politics, at least during the Republican period. In order to hold a military command, an individual will have had to achieve electoral success, and whatever one's view regarding the much-debated balance between aristocratic and popular power in Republican political life, this will have required an individual to be

[6] For examples where personal intervention worked, see Plut. *Mar.* 20 (with Goldsworthy 1996: 154–5), Front. *Str.* 2.8.12 (Sulla), Cass. Dio 75.6.7 (Septimius Severus), Amm. Marc. 31.10.13 (Gratian); and where intervention led to wounding or worse, with wider adverse consequences, see Livy 25.34 (Publius Scipio), Joseph. *BJ* 3.236–8 (Vespasian), Tac. *Hist.* 4.34 (Civilis). For discussion of a general's options as to where he positioned himself during battle, see Goldsworthy 1996: 149–63.

[7] Cf. also Agricola at the battle of Mons Graupius (Tac. *Agr.* 35.4). The value of such a gesture had to weighed against the disadvantages of the general being less visible to his men, less mobile, and less able to survey the battlefield (Goldsworthy 1996: 151).

assertive and a competent speaker, whether in meetings of the senate or of citizens (*contiones*). Lucullus, e.g., was said to have been equally clever at speaking in the forum and in military camps (Plut. *Luc.* 33.3). It is surely also significant that individuals who acquired reputations for successful military leadership during the Republic tended to be those who, like Pompey, held extended periods of command which facilitated their learning by experience. The protracted Hannibalic War offered Scipio Africanus his opportunity, just as Marius benefited from the prolonged Germanic threat to Italy at the end of the second century BC, and Caesar from the chance to extend his time in Gaul. Such lengthy commands, however, were the exception, going as they did against the Republican ethos of limiting tenures so as to provide equal opportunities for military glory among the elite. It is less easy to follow this strand through the Principate, since the emperor was regarded as having an inherent monopoly on *auctoritas*, etymologically related as it was to the title of Augustus. In Late Antiquity, on the other hand, emperors of the later third and fourth centuries tended to be men of military experience and competence who would not have risen to political prominence without a reputation for leadership in the military field. Although emperors of the fifth and sixth centuries usually did not campaign in person, they sometimes had military backgrounds prior to their accession, and their generals had usually pursued military careers, with a number in the fifth century (e.g., Stilicho, Aetius, Aspar) able to exert considerable political influence on the strength of their military reputation.

The fourth and final quality of 'good fortune' may seem even less easy to generalise from, but in fact good generals could help to make their own luck. 'It is not surprising that . . . the morale of troops should have been stimulated by the belief that their general had an uncanny knack of winning his battles, and that he owed this success to some supernatural agency with which he was privileged to be in very particular and intimate contact.'[8] So, e.g., Scipio Africanus 'strengthened the confidence of the men under his command and their readiness to face dangerous enterprises by instilling in them the faith that his plans were divinely inspired',[9] as when he told his troops that his audacious plan to attack New Carthage in Spain across the tidal shallows of the adjacent lagoon in 209 BC had been

[8] Balsdon 1951: 2, with further discussion in Kragelund 2001: 75–95, Harris 2009.

[9] Polyb. 10.2.5, although Polybius then emphasises that 'his actions were invariably governed by calculation and foresight' (no doubt mindful of Scipio's critics who 'thought that luck played the chief part in his successes': Balsdon 1951: 2).

suggested to him in a dream by, appropriately enough, the sea god Neptune.[10] In the uncertain circumstances of his decision to march on Rome with his army in 88 BC, Sulla claimed to have had a dream in which a female deity gave him a thunderbolt with which to strike his enemies (Plut. *Sull.* 9), and Sertorius was said to have strengthened the resolve of his troops through various devices, including dreams and specious religious signs (the most famous being his pet white deer, through whom he claimed the goddess Diana communicated in his sleep)(Aul. Gell. 15.22, Plut. *Sert.* 11). Lucullus received reassurance from Aphrodite in a dream prior to a (successful) naval battle with Mithridates (Plut. *Luc.* 12), while Caesar claimed to have seen a divine being encouraging him to cross the Rubicon in 49 BC (Suet. *Iul.* 32). Augustus put dreams to a somewhat different use in the context of battle, explaining his apparent flight at Philippi in 42 BC as a case of his cheating death through divine intervention via a dream (Plut. *Brut.* 41).

With the advent of monarchy, any suggestion of divine favour towards an individual outside the imperial family was risky, so that the main instances during the Principate relate to individuals successfully challenging for power in the context of civil war – Vespasian's experiences in the temple of Serapis in 69 and Septimius Severus' dreams, such as his rival Pertinax being thrown from his horse[11] – and these were obviously as much about securing wider political legitimacy as they were about strengthening soldiers' morale on the eve of a battle.[12] Constantine can be seen combining these two traditions in the early fourth century through his dream/vision of the heavenly sign from the Christian God on the eve of his battle against Maxentius outside Rome (312), having previously claimed to have had a vision of the god Apollo in Gaul, promising him victory.[13] The following year, on the eve of his battle against Maximinus Daia, his fellow emperor Licinius reported a dream in which an angel of God had given him the words for a pre-battle prayer to be used by his troops (Lactant. *De mort. pers.* 46). Continuing in the context of civil war, Constantine's son Constantius II was assured of victory against the usurper Magnentius at Mursa in 351 by the local bishop relaying an angelic

[10] Polyb. 10.11.7. See Richardson 2018 for a critical re-evaluation of this episode.
[11] Suet. *Vesp.* 7, Cass. Dio 75.3.
[12] Note, however, Germanicus' resorting to a favourable dream to encourage his men on the eve of battle against Arminius in AD 16 (Tac. *Ann.* 2.14).
[13] For the variant traditions, see Lactant. *De mort. pers.* 44, Euseb. *Vit. Const.* 1.28–32; vision of Apollo: *Pan. Lat.* 6.21. For an attempt to reconcile all three accounts, see Weiss 2003, with the sceptical response of Harris 2005.

message, while Theodosius I spent the night in prayer after a difficult first day in the battle of Frigidus against the usurper Eugenius in 394, until he had a vision of two saints assuring him of victory the next day.[14] It is understandable that some scholars have seen the reports of these later episodes as the result of *post eventum* elaboration by Christian authors, but given the long tradition of such reports of divine aid and direction as a strategy for boosting morale and strengthening confidence in a commander, it is plausible that Constantius and Theodosius made use of them in the context of two very hard fought battles. An interesting variation on divine help emerged in the late sixth century when the general Philippicus paraded a sacred icon of Christ through the ranks prior to the battle of Solachon in 586, 'thereby inspiring the army with a greater and irresistible courage' (Theoph. Sim. 2.2.4–6).

Exploitation of divine signs presupposes some form of communication with the troops. This could have been achieved by initiating appropriate rumours, but it seems more likely that generals will have announced them directly through speeches to officers and/or men prior to battle. The subject of pre-battle speeches is, however, a contentious one, for two reasons.[15] First, there is understandable scepticism that the speeches attributed to generals in ancient histories can be accurate verbatim records of what was actually said (who of those present would have had the inclination to keep a written record, and would ordinary soldiers really have appreciated the rhetorical tropes and literary allusions which feature in many examples?); and secondly, there is also understandable scepticism about the practicalities of a general making himself heard by more than a small proportion of his troops. On the first of these issues, there seems little doubt that the authors of histories in classical antiquity saw the pre-battle speech (ideally in pairs – one for each side) as an opportunity to display their rhetorical prowess by placing in the mouths of opposing generals sentiments that seemed appropriate to the occasion, but had only a very loose relationship to reality. In all this, there looms large the precedent of Thucydides who famously admitted that he used a strong element of authorial invention in the speeches (by no means all pre-battle) in his history of the Peloponnesian war (1.22.1–2). There is, however, an item of evidence that appears not to have been adduced in relation to this subject. In the course of its account of the Roman-Persian war of the early sixth century, the *Chronicle* traditionally

[14] Sulp. Sev. *Chron.* 2.38, 5, Theod. *Hist. eccl.* 5.24.

[15] The debate was initiated by Hansen 1993; subsequent discussions include Goldsworthy 1996: 145–8, Hansen 1998, Anson 2010, Lendon 2017: 145–54.

attributed to Joshua the Stylite includes a short speech by the Roman general Patricius to his troops when his army found itself cornered by Persian forces, with their backs to a swiftly flowing river in flood which swept away those mounted troops who ventured to cross it:

> When Patricius saw this, he encouraged the soldiers, saying, 'Romans, let us not disgrace our race or our military calling by fleeing from our enemies, but let us turn around and fight them, for we may have the measure of them. But even if they are too powerful for us, it is better to die by the edge of the sword with a good reputation for courage than to drown like cowards in a torrent of water.' Then the Romans, because the river gave them no choice, were persuaded by his exhortation and furiously turning against the Persians, destroyed them, capturing their leaders alive (Josh. Styl. 66 [tr. Trombley & Watt]).

What is significant about this speech is that it appears in a text from outside the usual cultural context, written, not in Latin or Greek, but in Syriac, and therefore providing an insight into pre-battle speeches less encumbered by the traditions of classical historiography. This is not to suggest that Patricius addressed his men in Syriac, rather than Greek or Latin, but if the *Chronicle*'s account of his speech is more likely to reflect what a commander actually said, then what is notable is its appeal to honour and courage, its practical good sense, and its brevity (this last perhaps due to the need to act swiftly in these particular circumstances). None of this is surprising, but it is an item of evidence which deserves consideration in the context of this longstanding debate.

As for the second issue of practicalities – the general making himself heard – all manner of comparanda have been adduced in attempts to clarify matters. In the process, a range of possibilities has emerged, both temporal and spatial. It has been pointed out that pre-battle speeches need not necessarily have been given immediately before engagement, but could be delivered earlier in the day or even on the day before. These alternative possibilities in turn have made it more feasible that a general would have time to disseminate his sentiments throughout the army by officers addressing individual units. Even in the case of troops being addressed immediately prior to combat, it has been recognised that this could take the form of the general repeating short phrases as he passed along the lines on horseback. Gesture – an important aspect of rhetorical delivery – could also play a role in communicating with troops.[16] To the diverse material

[16] See, e.g., Suet. *Iul.* 33 (also illustrating the scope for misinterpretation). For rhetorical gestures, see Quint. *Inst.* 11.3.85–116.

which has featured in discussions can be added one further item which articulates some of these points more explicitly than much of the evidence but does not appear to have been deployed previously, namely the advice in the handbook for generals attributed to the late sixth-century emperor Maurice (the *Strategikon*). In adducing this material, it is worth bearing in mind that this treatise 'was undoubtedly an official ordinance, sponsored by central government rather than the personal and/or amateur reflections which in large part characterise this broad genre'.[17] Guidance for the general is divided into two categories – that for the day before battle, and that on the day of the battle itself. Significantly, there is a section on 'Using speeches to encourage the troops' which appears, moreover, in the first of these two categories, and is explicit that the general should do so 'during a rest period' and should not try to address all the troops in one place, but do so 'by *meros* or *moira*' (a *moira* comprised between one and two thousand men, and a *meros* three *moirai*). As for content, speeches should 'recall their previous victories and promise rewards from the emperor and recompense for their loyal service to the state' (*Strat.* 7A.4).

As much as the words the general spoke, his visible presence was an important part of his leadership.[18] Allied to this was the emergence of the idea of the commander as fellow-soldier (*commilito*). Even when he did not engage directly in fighting, it was possible for a general to claim such identity particularly through commensality – sharing the same food as the rank and file – and participating in their training and labours. Cato the Elder and Marius – both 'new men' from less aristocratic backgrounds – seem to have been the first generals to have made a point of doing so, with the pressure to identify with one's troops in this way growing as military support became increasingly crucial during the Late Republic.[19] After the end of the civil wars, Augustus famously made a point of refusing to address soldiers as *commilitones* in public contexts, in order to emphasise his independence from the armed forces (Suet. *Aug.* 25.1), but this usage, and symbolic gestures of identification, gradually resumed during the first century, particularly after the insecurities of the civil wars of 68–69. Thereafter most emperors made at least some sort of gesture in this direction, with the civil wars of the 190s prompting greater emphasis during the Severan dynasty – until another man from a less aristocratic background (Maximinus) was able to seize power in 235, partly on the strength of his military career giving substance to his claims to be a

[17] Rance 2004a: 267. [18] Cf. Anson 2010: 316–17. [19] Phang 2008: 276–7.

commilito.[20] With emperors in the later third and fourth century increasingly emerging from the military and commanding in person, use of this language of identification became natural, both in descriptions of emperors in panegyric and in official pronouncements directed to troops. Although it might have seemed inappropriate to continue with this after the abandonment of personal involvement in campaigning by emperors after 395, the usage of the language of identification seems, if anything, to have intensified, with emperors and their advisers appreciating the need to try to bridge the gap between palace and military camp.[21] Of course emperors' assertion of their status as fellow-soldiers arose as much, if not more, from concern about the political loyalties of troops as it did about issues of morale in battle. However, late Roman generals are also presented as using this language of identification in specifically military contexts – so, Theodosius the Elder during his north African campaign in 375, and Belisarius during the Vandal and Gothic wars of the 530s[22] – suggesting enduring recognition of the value of this strategy on the part of astute generals.

4.2 Obedience and Mutiny

Obedience is of course a fundamental requirement on the part of soldiers in any army, above all when it comes to confronting the possibility of death or serious injury in battle. But such are the stakes in that context, and the natural reluctance of individuals to expose themselves to such risks, that it is important to make obedience and discipline a broader part of military culture. In modern and early modern armies this culture has been inculcated through such strategies are as a focus on drill and on the de-emphasising of individuality through the wearing of uniforms. Another facet of obedience is the wider issue of ensuring the loyalty of armed forces to the existing political regime.

In the Roman context, drill and the wearing of uniforms seem not to have been viewed as so important in instilling conformity.[23] However, other aspects of Roman practice, certainly not unique to this context, can be identified as important. One obvious ritual was that of swearing an oath

[20] Campbell 1984: 32–59.

[21] On the language of identification in official pronouncements, see Lee 2007: 61–6; for emperors sharing the conditions of ordinary soldiers in panegyric, see Blockley 1975: 87–8.

[22] Theodosius: *contubernales devoti* (Amm. Marc. 29.5.22); Belisarius: *andres sustratiotai* (*Proc. Bell.* 3.19.2, 4.15.16, 5.28.6).

[23] Phang 2008: chs. 2–3.

of obedience at the start of service.[24] During the Republic, an individual soldier will have done this a number of times during his life, since it was required of all those selected to serve each year (at least from the late third century BC onwards). Unsurprisingly, the civil wars of the first century BC posed acute questions about allegiances, so that generals frequently required troops to take an oath of loyalty to them, either when taking over command of units or at some other point.[25]

These more personal oaths of allegiance were perhaps the context out of which arose Augustus' use of oaths to himself (and not just by soldiers but also civilians). At any rate, the military oath must have been modified with the advent of emperors so as to incorporate an expression of loyalty to the current ruler, with annual renewal on 3 January. The precise formula used during the Principate has not been preserved, but there are numerous references to the military oath being sworn, while potentially relevant detail has been preserved in a text from the late fourth or early fifth century:

> They swear by God, Christ and the Holy Spirit, and by the majesty of the emperor which second to God is to be loved and worshipped by the human race . . . The soldiers swear that they will strenuously do all that the emperor may command, will never desert the service, nor refuse to die for the Roman state (Veg. *Mil.* 2.5 [tr. N. Milner]).

This text shows that by Late Antiquity an explicitly Christian dimension had been incorporated into the oath, but it also provides some idea of what the oath probably involved in earlier centuries, above all the emphasis on loyalty to the emperor. Because of this element, the oath appears to have been renewed annually on the accession date of the current emperor, a practice which will have served to reinforce it. Clearly the oath did not prevent some soldiers from deserting or from turning against the reigning emperor, but neither should its value be underestimated. As the usual term for it, *sacramentum*, implies, it was seen as involving religious sanctions, and one ancient commentator claimed that the Romans observed the military oath more than any other (Dion. Hal. *Ant. Rom.* 11.43). The importance attached by soldiers to the religious dimension of the oath is corroborated by the evidence from the Principate for soldiers honouring

[24] Discussion of the oath in Campbell 1984: 19–32, Lee 2007: 52–3, Phang 2008: 117–20.

[25] E.g., Plut. *Sull.* 27, App. *B. Civ.* 1.66, 4.62, 4.116, Caes. *B. Civ.* 1.76, 2.28–32, 3.87. As John Rich notes (pers. comm.), the Caesar passages all relate to the Pompeians, with Caesar appearing to imply that he himself did not need to do this.

the *genius* – the guardian spirit – of the oath.[26] It has been argued that even in modern times the military oath carries significant symbolic force and should not be viewed as a piece of empty ritual.[27]

The influence of the oath on soldiers' behaviour was reinforced by a range of other strategies that provided positive and negative incentives for compliance. In the positive category, the role of rewards for bravery in encouraging displays of *virtus* has already been noted in a previous chapter (Section 2.2), nor should the motivational impetus provided by the promise of a share of booty be underestimated. The prospect of discharge benefits, whether in the form of land or cash, must also have played an important part in persuading soldiers to complete their period of service in good standing.

In the negative category falls the evidence for punishments. These are portrayed in very stark and uncompromising terms by Polybius, writing of the army in the second century BC. He refers to the use of decimation – the execution of 10 per cent of the men in a unit – and cudgelling, alongside beating and ritual humiliation, as penalties for a range of offences, some seemingly quite trivial (6.37–8). Other sources also retail accounts of capital punishment for individuals who contravened orders, even when their actions involved displays of heroism which benefited the state. The latter stories are redolent of rhetorical exaggeration which must cast doubt on their historicity, but even if the measures described by Polybius really were implemented with the severity he indicates during the middle Republican period, then it seems that this was not maintained to the same degree in later periods of Roman history, at least in relation to the most severe and seemingly arbitrary punishments.[28] It is evident from the military mutinies which followed the death of the emperor Augustus in AD 14 that excessive use of beatings could eventually encourage an unhelpful level of disaffection which was not ultimately in the interests of the Roman state. It has been argued instead that discipline was inculcated primarily through troops engaging in physical labour on relevant projects such as camp construction.[29]

However, despite measures to encourage discipline and compliance on the part of troops, there were occasions when soldiers were unwilling to

[26] Apul. *Met.* 9.41, *AE* 1924.135. [27] Holmes 1985: 32–3.

[28] Decimation was not, however, completely unknown during the Principate (Suet. *Aug.* 24, Tac. *Ann.* 3.21, Suet. *Galb.* 12), an episode involving a version of decimation is reported in the fourth century (Amm. Marc. 24.3.1–2), and it was still a prescribed penalty in the late sixth century (Maur. *Strat.* 1.8.17).

[29] Phang 2008: chs. 4, 6.

conform. At an individual level, this could take the form of desertion, whether failure to return from leave or simply slipping away from camp. It is clear from discussions by jurists that desertion was an issue of concern during the Principate, and the authorities tried to exercise tight control over the granting of leave.[30] However, losing individual soldiers in this way was less problematic than whole units refusing to obey orders. In discussing the incidence of mutiny, it is important that it be distinguished from civil war and usurpation. While the distinction is sometimes blurred, it was essentially a matter of initiative: mutiny generally arose from dissatisfaction among the rank and file which led them to challenge the authority of their general and officers, whereas civil war generally arose from generals persuading troops to support them in a bid for political power (on which, see further in Section 4.3).

In considering the subject of mutiny, it is important to appreciate that it sits at one end of a spectrum of ways in which soldiers might express their dissatisfaction, alongside insubordination and expressions of grievance.[31] Mutiny did not usually begin without soldiers first voicing their concerns about issues, whether relating to military strategy or conditions of service. Those concerns might be conveyed to the commander through intermediaries such as centurions and officers or direct to the commander in camp meetings, which gave commanders an opportunity to respond, whether by reassuring his troops or indicating his intention to modify his plans.[32] Often that sufficed to defuse tensions, but when commanders were slow to respond or did not offer adequate concessions, soldiers' disaffection could spill over into open mutiny.

Modern discussions offer varying counts of mutinies during the Republic – one study refers to forty instances, another to sixty.[33] These variations arise from differing definitions of mutiny and differing assessments of the historicity of episodes from the early Republic. There is, however, agreement that mutiny was a more common phenomenon during the Republic than one might assume on the basis of the popular image of a thoroughly disciplined Roman army or its overall record of military success, and that there was a particular concentration of episodes during the first century BC. The instance from before the first century BC about which the sources preserve the most detail is the mutiny in 206 BC during the

30 *Dig.* 49.5; Speidel 1985.
31 For issue of terminology and categorisation, see Brice 2015.
32 Chrissanthos 2004.
33 Forty: Messer 1920 (my count of the episodes to which he refers, which finishes in 30 BC); sixty: Chrissanthos 2013: 324 (also covering to 30 BC, presumably drawing in part on his 1999 PhD thesis, which lists forty-six instances down to 40 BC).

Hannibalic War by some of Scipio Africanus' troops at Sucro in Spain.[34] Although circumstantial factors such as Scipio falling ill contributed, the underlying causes were four issues which, in varying combinations, often recur in the complaints of soldiers in subsequent episodes – delays in pay, dissatisfaction with the division of booty, length of service, and shortages of supplies. By way of illustration, division of booty, length of service and shortages of supplies were also factors in the mutiny against Lucullus in Asia in 68 BC, along with difficult winter conditions (Plut. *Luc.* 32–5, Cass. Dio 62.6), while that against Caesar in 58 BC at Vesontio in Gaul was attributed (by Caesar) to food shortages and fear of an unfamiliar enemy (*B. Gall.* 1.39–41). A further mutiny against Caesar in 49 BC at Placentia arose from delays in pay, shortages of supplies, and a prohibition on plundering, with similar factors at work in yet another mutiny against Caesar in 47 BC, together with length of service.[35] Dissatisfaction with conditions of service was, then, the common theme in mutinies during the Republic.[36] As for how commanders responded, there were episodes where punishment was inflicted, at least on those regarded as ringleaders – so, Scipio at Sucro and Caesar at Placentia.[37] However, it was more common for concessions to be made, even if accounts sometimes try to mask this in their concern to defend the reputation of the relevant commander.[38]

During the Principate, there is much less evidence for dissatisfaction and mutiny.[39] Given that discontent with conditions of service was the common theme in Republican mutinies, a decrease in the frequency of mutiny during the Principate is most easily related to Augustus' establishment of a standing army with regular pay and discharge bonuses. Those changes did not, of course, take place until well into Augustus' reign,[40] nor

[34] Chrissanthos 1997.

[35] Details and discussion in Chrissanthos 2001: 67–70. Many instances of ill-discipline during the civil wars of the late 40s and 30s BC were deliberately provoked by opposing generals trying to win over manpower (see further Brice 2011).

[36] Fulkerson (2013: 165) notes that the predominantly negative attitude of ancient sources towards Roman soldiers 'renders all the more noteworthy their willingness to record grounds for the soldiers' complaints, even when they suggest they are mere pretence.'

[37] Fulkerson (2013: 165) suggests that being able to lay the blame on a small number of individuals was 'extremely convenient for generals, who would prefer to punish the fewest number of soldiers possible while maintaining discipline.'

[38] Brice (2015) argues, *pace* Chrissanthos 2001, that Caesar did punish the mutineers of 47 BC since they were discharged with less than the promised benefits.

[39] Brice 2003 argues that there is more evidence for military unrest during the Julio-Claudian period than is usually recognised (e.g., the initial refusal of troops to cross the Channel in 43: Cass. Dio 60.19.2–3), but still concedes that examples of soldier-initiated unrest are limited.

[40] Brice (2003: 269–73, 283) notes instances of unrest during campaigns in Spain in the mid-20s and 19 BC.

did they prevent major mutinies in Pannonia and on the Rhine following Augustus' death in AD 14, where the chief grievances were length of service and pay, along with complaints about excessively harsh discipline (Tac. *Ann.* 1.16–71) – with the exceptional circumstances of the first succession under the new regime of the Principate providing the occasion. 'In AD 14 . . . the system was still not working smoothly. Men were being retained long beyond the official maximum for service, presumably because the money was not available to pay their gratuities.'[41] However, since episodes like these did not recur during the Principate with the regularity they had during the Republic, it looks as if the new system gradually bedded in, even if the level of pay was hardly generous (Section 3.2).

Mutinous soldiers are sometimes seen as a common feature of the empire's turmoil in the mid-third century, but it is difficult to identify episodes that can be classed as mutiny rather than civil war arising from usurpation. There were undoubtedly instances when troops killed their commander emperor, but this was usually in the context of confronting a challenger for the throne to whom, for whatever reason, the troops had decided to switch their allegiance. One episode that is closer to Republican precedents was the mutiny against Maximinus at Aquileia in 238 in which food shortages seem to have played a critical role in turning the troops against him (Herodian 8.5), while according to some sources, Probus' downfall in 282 was the result of his troops mutinying after he required them to undertake marsh-drainage (Aur. Vict. *Caes.* 37.4, Eutr. 9.17.3). Overall, however, evidence for mutinies in Late Antiquity is limited until the sixth century.[42]

In 536, early in the reign of Justinian, a serious mutiny broke out among troops who had participated in the campaign to recover north Africa from the Vandals and who then remained there on garrison duty. Delay in the payment of troops contributed (*Proc. Bell.* 4.15.55), but other factors do not have obvious comparanda from the Republic. One was the the imperial government's decision to return land seized by the Vandals in the fifth century to the descendants of its previous owners. This created

41 Dobson 1986: 14.

42 Julian's proclamation as emperor by his troops in Paris in 360 is sometimes referred to as a mutiny, but there is good reason to regard this as a case of usurpation in which Julian and his allies exploited the discontent of soldiers to their own ends (Bowersock 1978: 46–52). A few years earlier, Julian had praised Constantius II for ensuring that his troops were adequately supplied, so that 'they were not driven to insubordination by lack of necessities' (*Or.* 1.22A; cf. *Or.* 2.88A–B). For possible but inconclusive instances of mutiny in the fifth century, see Marcellin. *Chron. s.a.* 420, Malch. *frr.* 18,3, 27.

disaffection because many of the imperial troops married Vandal women who stood to lose out from this decision. Secondly, church services which used the heterodox Arian liturgy were banned. Although this was directed at the residual Vandal population, some of the imperial troops who had been recruited from barbarian groups were also Arian, so were affected too.[43] This mutiny dragged on for a decade, with the government deploying a combination of concessions and force in an effort to terminate it. However, further episodes in north Africa and Italy during this period showed ongoing discontent by soldiers at delays in receiving their pay[44] – delays which were largely a result of the multiple military commitments which the Roman state faced in this period, both in north Africa and further afield.

Further instances of mutiny from the sixth century had much in common with the mutinies of the Republican period. The first of these involved military units on the eastern frontier where, in 579, a large number of troops refused to fight until they received their pay in full. The emperor Tiberius II responded by sending a senior court official with sufficient gold to appease the men, who then agreed to return to their posts (Ioh. Eph. *Hist. eccl.* 6.28). In 588, news that part of their pay would be in kind rather than coin and there was to be a change in the length of service (presumably to longer) triggered a mutiny in the same region which lasted for twelve months until a local bishop mediated. The troops were granted an amnesty and it seems that the proposed changes were not introduced.[45] The most serious mutiny, however, involved the Danube army in 602, which led to the overthrow of the emperor Maurice and an extended period of very damaging upheaval for the empire.[46] This episode did not emerge without forewarning, for the Danube army had shown mutinous tendencies during the 590s, arising from discontent with their conditions of service, but commanders had diffused these situations through persuasion or concessions (Theoph. Sim. 6.7–8, 6.10.1–3, 7.1).

The mutiny of 602 was precipitated by the emperor Maurice's (renewed) insistence that units winter north of the Danube to combat the threat from Slavic raiders. There may have been sound reasons for this

[43] Proc. *Bell.* 4.14.8–15. This was despite the fact that they had enlisted with legal exemption to remain Arian (*Cod. Iust.* 1.5.12 (17) [527]).

[44] *Proc. Bell.* 4.18.9, 4.26.10, 7.11.13–16, 7.12.7–8, with discussion in Kaegi 1981: 47–51.

[45] Theoph. Sim. 3.1–3, Evag. *Hist. eccl.* 6.4, 11–13; Kaegi 1981: 68–72, Whitby 1988: 286–9.

[46] Discussions in Kaegi 1981: ch. 5, Whitby 1988: 165–9.

strategy – it was easier to avoid Slav ambushes and to force them to give battle when forests were bare and rivers frozen – but the troops were understandably less than happy about having to campaign in difficult and inhospitable conditions, as Lucullus' troops had been in Asia in 68 BC. Mirroring other Republican issues, there were also concerns about booty (Theoph. Sim. 8.6.2) and about pay and supplies (Ioh. Nik. 102.10). As soldiers had often done during the Republic, these troops communicated their concerns to their commander Peter through representatives. Unlike his predecessors in the 590s, however, Peter refused to make any concessions – but neither did he act decisively against the ringleaders of the mutiny in the way that Scipio had at the Sucro and Caesar at Placentia, which proved a fatal error. The troops soon organised themselves to march on Constantinople where they proceeded to overthrow Maurice, with disastrous consequences for the empire.[47]

Overall, then, the incidence of military mutiny in Roman armies was quite variable. There were particular concentrations in the late Republic period and the sixth and early seventh centuries, with discontent over conditions of service the predominant theme. Augustus' establishment of a standing army with fixed conditions of service appears, after the initial setback of the mutinies of AD 14, to have largely dealt with the issues which encouraged mutiny during the late Republican period. However, the demands of warfare on multiple fronts during the sixth century placed strains on the imperial treasury which resulted in delays in pay and contributed to another period of frequent mutiny, with the additional issue of over-wintering north of the Danube precipitating a mutiny of momentous consequences in 602. Interestingly the overwhelming trend, whether during the Republic or Late Antiquity, was for the authorities to accede to the demands of the troops. While the events of 602 provide a stark warning of the serious consequences that could ensue when there was a refusal to take heed of soldiers' concerns, this refusal was particularly unwise because of the relative geographical proximity of the troops to the imperial capital, whereas the north African mutiny in 536 – and indeed those of the Pannonian and Rhine legions in AD 14 – was less likely to pose a direct threat to the emperor's position because of their respective distances from Constantinople and Rome.

[47] Although Maurice's replacement by the officer Phocas might make this seem like a case of usurpation, there is no suggestion in the sources that Phocas initiated the mutiny.

4.3 The Dilemmas of Civil War

Simply in terms of the amount of time it occupied and resources it consumed, civil war was an important category of warfare in the Roman world, even if it was not as frequent as warfare against non-Romans. During the Republic, instances of civil war were concentrated in the final half century or so of its existence, beginning with Sulla's conflict with Marius in the 80s BC and ending with Octavian's war with Antony in the late 30s. During the Principate, there were two major episodes, in 68–69 and 193–7; the lengthy gap from the end of the Republic until 68–69 perhaps reflected a determination to avoid a recurrence of the miseries of civil war, and likewise the even longer gap after 68–69.[48] The mid-third century was another concentrated period, while there were periodic bouts across the fourth century, starting with Constantine's rise to power and ending with Theodosius I fending off challenges in the west. The fifth century witnessed further instances in the early decades, especially in the west, and finished with that of Anastasius against the Isaurians in the 490s. The sixth century, on the other hand, was remarkably free of civil war, apart from Vitalian's challenge to Anastasius in the second decade.[49]

The high casualties sometimes resulting from civil war also make this a significant phenomenon. Of course the casualty figures preserved by ancient sources need to be treated with caution, because of the potential for authors to exaggerate if it suited their purposes, as ancient historians themselves recognised with particular reference to civil war (cf. Herodian 3.7.6). Nevertheless, it is evident that some battles resulted in large losses of life, even if there must remain some doubt about the specific figures. The battle at Philippi between the forces of Antony and Octavian on the one hand and those of Brutus and Cassius on the other (42 BC) may have involved losses of 20,000 men on each side;[50] the second battle of Cremona between the Vitellian and Flavian forces (69), together with civilians killed in the city's subsequent sack, is said by one source to have resulted in 50,000 deaths (Cass. Dio 65.15.2); the same source gives

[48] For a helpful overview of the major instances of civil war in the late Republic and Principate, see Osgood 2015. The abortive rebellion of Scribonianus against Claudius in 42, although short-lived, was viewed as sufficiently serious to warrant the erasure of his name from public monuments (Levick 1990: 59–60), which may suggest a greater sense of insecurity on the part of the Julio-Claudian emperors than appears in retrospect.

[49] For a catalogue of civil wars in Late Antiquity, see Drinkwater and Lee 2014: 213–14. For a catalogue of usurpations from 27 BC to 455, see Omissi 2018: 20, 313–15. For a possible explanation of the absence of civil war for most of the sixth century, see Lee 2007: 71–2.

[50] Brunt 1987: 487–8, who likens it to the slaughters at Cannae and Arausio.

20,000 dead on the losing side in the battle of Issus (194) between the armies of Septimius Severus and Pescennius Niger (74.8); and both sides (those of Constantius II and Magnentius) are reported as having lost at least half their forces in the battle of Mursa (351) (Zonar. 13.8) – forces, as one near-contemporary source ruefully observed, 'adequate for any number of foreign wars, which might have provided many a triumph and much security' (Eutr. 10.12).

As important as the quantitative dimensions, however, was the way in which civil war turned Roman warfare on its head by pitting Roman against Roman. The negative image of civil war has already been noted (Section 1.1), and although this image was to some extent the creation of victorious regimes keen to gain credit for having ended civil war and restored stability, this cannot have been the whole story. It is understandable that there should have been a revulsion against civil war on the part of non-combatants affected by it. But even soldiers must surely have found the prospect of confronting another army which also proclaimed its allegiance to the Roman state an unnerving prospect, at least on the first occasion. The Roman ideology of victory was based in part on the conviction that the gods or God favoured the Roman cause in war, so how did one rationalise a situation where the enemy claimed the favour of the same gods or God? Moreover there was a well-known anecdote from the first period of civil war in the Republic in which a soldier fighting on one side discovered that he had killed his own brother fighting for the opposition, and in his grief committed suicide[51] – thereby exemplifying the worst potential consequences of civil war, as well as symbolising the way it transgressed natural social bonds. A similar sentiment was expressed by Caesar when justifying his decision in 49 BC to starve enemy forces in Spain into surrender: he was reluctant to see fellow-citizens suffer unnecessary bloodshed (Caes. *B. Civ.* 1.72). Likewise, the historian Cassius Dio imagined the two sides at Pharsalus (48 BC) as reluctant to engage: 'Sprung from the same country and from the same hearth, with almost identical weapons and similar formation, each side recoiled from beginning the battle, recoiled from killing anyone' (41.58).[52] If, then, civil war was in some sense seen as unnatural or counter-intuitive, at least in this period,

[51] Livy *Per.* 79, Val. Max. 5.5.4, Tac. *Hist.* 3.51, Granius Licinianus 35.18, Oros. 5.19.12–13. The context was fighting in Rome in 87 BC between the forces of Cinna and Pompeius Strabo.

[52] Cf. Cass. Dio 65.13 on the second battle of Cremona (69) where he presents soldiers of the two sides conversing across the lines during the night and even sharing food.

why were Roman soldiers nonetheless prepared to engage in it repeatedly throughout Roman history?

Given that it was an unprecedented phenomenon at this stage of Roman history, it is unsurprising that late Republican leaders felt the need to justify the decision to embark on civil war to their troops. In 88 BC, while indicating his reluctance to resort to force, Sulla emphasised the violence threatened against his person, as consul, by his enemies, Marius and Sulpicius, with the implication that the latter were violating constitutional principles. Perhaps even more importantly he repeatedly articulated what was to become an important refrain until the end of the Republic – his concern to free the state from tyrants (App. *B. Civ.* 1.57). In a similar vein, Caesar justified his decision to invade Italy in 49 BC as arising from his desire to 'restore the Roman people to *libertas* from their oppression by a faction of the few' (*B. Civ.* 1.22.5). Not content with quasi-constitutional justifications, however, Sulla sought to strengthen his hand in 88 BC by the further strategies of claiming to have had a dream (Caesar also reported a vision at the Rubicon: see Section 3.1) and by disseminating a rumour that if Marius was allowed to take Sulla's military command against Mithridates, he would enlist other troops and so deprive Sulla's men of the valuable booty expected from the campaign. These various considerations were not enough to induce Sulla's officers to support him in what they regarded as an attack on their homeland – a telling indication of just how momentous a step this was – but they were sufficient to persuade the rank-and-file, who proceeded to march on Rome enthusiastically. As has often been noted, in 49, by contrast, all but one of Caesar's officers backed him, implying that attitudes to civil war had begun to change in the intervening decades.

Quasi-constitutional justifications continued to be important in the civil wars following Caesar's death, not least because generals had become aware of the fragility of loyalties and the potential for desertion (cf. App. *B. Civ.* 5.17). During 44–43, Octavian and his allies justified their military action against Antony by reiterating the claim to be defending the *libertas* of the state against oppression[53] – an argument subsequently monumentalised in the opening lines of Augustus' *Res Gestae* (1.1). In his conflict with Antony in the later 30s, Octavian developed further arguments, including the powerful allegation that Antony intended to relocate the capital of the empire from Rome to Alexandria.[54] Alongside these ideological incentives the competitors for power following Caesar's death made extensive use of

[53] Arena 2012: 263. [54] Osgood 2006: 353–4.

promises of substantial cash donatives to maintain loyalty of troops to their cause, while a primary inducement for soldiers who had served under Caesar was their concern to ensure that the land allotments they had received from Caesar were not revoked.[55]

The prospect of booty continued to be an incentive in the civil war of 68–69,[56] but also important was the portrayal of the opposition as non-Romans. In the face of the Vitellian invasion of northern Italy, the emperor Otho presented Vitellius' forces as comprising Germans and native tribes from Gaul and the Rhine frontier – and indeed during the siege of Placentia, Otho's troops are said to have derided the Vitellian forces as foreigners and aliens, while the latter are presented as approaching Italy as though it was a foreign land (Tac. *Hist.* 1.84, 2.21, 2.12).[57] Tacitus' literary goals no doubt played a part in the development of this theme,[58] but it is an entirely plausible one for leaders to have emphasised in the context of the changing patterns of recruitment to the Roman armed forces. During the civil wars of the Late Republic the legions were still drawn from within Italy, whereas the intake of Italians fell dramatically during the first century, replaced by growing numbers of provincials, especially from the more peripheral regions of the empire (cf. Section 2.1). These provincials came from a much more diverse range of ethnic and cultural backgrounds, and since legions tended to recruit from their locality, it will have been easier for legions from one part of the empire during the Principate to regard those from other parts as foreigners.[59] To cite one well-known example from the civil war of 68–69, towards the end of the night-long second battle of Cremona the Third Legion was reported as greeting the rising sun with cheers, reflecting their longstanding stationing in Syria and presumed adoption of some sort of solar worship (Tac. *Hist.* 3.24). All these legionary soldiers were of course Roman citizens, whether of long standing or more recent grant upon enlistment, but many would never have set eyes on the city of Rome, and the changing and ever more complex make-up of the legions shows that Roman identity itself was increasingly difficult to define. To these considerations must be added

[55] App. *B. Civ.* 2.125, 135, Nic. Dam. F130, 103. Further discussion of this and other issues in Brunt 1962: 75–84, Keppie 1983: 38–43 (including tabulation of cash donatives from 59–29 BC).

[56] Ash 1999: 37–8.

[57] There may be a hint of such a strategy already at work in 49 BC in the rumour that Caesar's army mostly comprised barbarians (Cass. Dio 41.8.6), while Octavian emphasised Antony's association with alien Egypt through Cleopatra.

[58] Cf. Ash 1999: 31–2, 39–50.

[59] Cf. Keppie 1997 (who, among other items, notes a second-century epitaph from Aquileia which contrasts service in the praetorian guard in Rome with service in a 'barbarian legion' [*ILS* 2671]).

another factor which differentiated the Principate from the Republic – the role of *auxilia* in the armed forces. Although units akin to *auxilia* contributed to the armies of the first century BC, they only became a significant component from Augustus onwards. Since these units were recruited from non-citizens, their presence in the armies of 68–69 must have helped to reinforce the sense that opposition forces were foreign.[60]

All this perhaps helps to explain why soldiers seem to have had fewer reservations about civil war in 68–69 compared with the Republican period. Tacitus is one of the sources for the anecdote cited above about a soldier unwittingly killing his brother during a bout of civil war in 87 BC, and then committing suicide when he realised what he had done. The historian refers to this story to provide a moral counterpoint to an episode from 69 when a soldier claimed to have killed his brother in a recent battle and then demanded a reward from his general (*Hist.* 3.51). He also refers to another episode from the same time when a son unwittingly killed his father. In this latter case, the son was duly penitent and conscious of the stigma of effectively being a parricide, prompting nearby troops to 'curse this cruellest of all wars.' However, Tacitus then continues to observe that 'this did not stop them killing and robbing relatives, kinsmen and brothers: they said to each other that a crime had been done – and in the same breath they did it themselves' (*Hist.* 3.25). Even after due allowance is made for Tacitean moralising,[61] these passages and the events of 68–69 imply fewer qualms about engaging in civil war on the part of soldiers. At the same time, Septimius Severus is still presented as appealing to principles of constitutional legitimacy when persuading his Pannonian troops to support his bid for power, arguing that Pertinax had been unjustly killed by the praetorians in Rome and his death needed to be avenged. He also portrayed the Syrian forces supporting Pescennius Niger as weakened by exposure to eastern luxury and not true soldiers or, by implication, true Romans (Herodian 2.10) – and there is also a possible echo of late Republican arguments in his claim to have 'restored the state (*res publica*)' in the inscription on his triumphal arch in Rome (*ILS* 425).

The civil wars of the Principate had developed out of situations in which the last emperor of a dynasty (Nero, Commodus) had alienated too many of the senatorial elite and also lacked a designated heir, with the eventual victor (Vespasian, Septimius Severus) managing to reign long enough to

[60] For the role of *auxilia* in 68–69, see Saddington 1982: ch. 9. For regional subcultures within the army, see Haynes 2013: 11, James 2014.
[61] And perhaps invention: Woodman 1998: 13–16.

ensure a stable transition to an adult son. That the death of Severus Alexander in 235 was not followed by a relatively short period of turmoil before a return to stability can be attributed in large part to the unprecedented external military challenges that the empire confronted from the emergence of Sasanian Persia to the east and more serious threats from Germanic groups to the north. As a result, the empire faced regular crises on one frontier or another, which generated political instability and opportunities for ambitious generals to make bids for power. However, the lack of detailed historical narratives for the mid-third century makes it difficult to determine how generals of this period motivated their troops to fight against other imperial armies. It is only in the fourth century that more detailed evidence about civil wars is again available. One theme from earlier periods of civil war that reappears is designating the other side as foreigners. The best example of this is from the usurpation of Magnentius in the early 350s, which 'one could not call a civil war, for its leader was a barbarian' (Julian *Or.* 1.42A), reflecting a concerted effort by the emperor Constantius II to portray Magnentius as a foreigner – despite the fact that he was almost certainly an ordinary provincial.[62] Likewise, the late-third century usurper Allectus was alleged to have barbarian troops and to have adopted their clothing and long hair (*Pan. Lat.* 8.16), while the early-sixth century rebel Vitalian was said by some sources to have been of barbarian descent,[63] perhaps again reflecting imperial propaganda. In the early fifth century west usurpers often sought support from barbarian groups increasingly present within the empire, thereby making themselves vulnerable to portrayal as foreigners, while at the end of the fifth century Anastasius' war against the Isaurians lent itself to presentation in such terms because of the Isaurians' reputation for brigandage, the hostility towards them which had grown during the reign of his (Isaurian) predecessor Zeno, and the fact that the imperial government was paying a very substantial annual subsidy to the region (whose welcome termination Anastasius' victory allowed).[64] What might be considered a variation on this strategy of presenting the opposition as 'the other' is the portrayal of Arbogast and his forces in 394 as pagans, fighting against the Christian Theodosius.[65]

Assertions of one's own constitutional legitimacy and the illegitimacy of one's opponent were clearly also important in the civil wars of the fourth century. Much of the most detailed extant evidence for this was produced after the event, but it must echo the propaganda strategies employed prior

[62] Drinkwater 2000; reservations in Omissi 2018: 174–5. [63] Haarer 2006: 164–50.
[64] Haarer 2006: ch. 2. [65] Cameron 2011: ch. 3.

to the decisive battles, whose target audiences included the relevant troops. One of those strategies was labelling one's opponent as a *tyrannus*, which in the fourth century had strong connotations of illegitimacy. It can be seen in the designation of Maxentius and Licinius as such in sources favourable to Constantine,[66] with Constantius II following the same pattern in relation to Magnentius (Julian *Or.* 1.1, 30B, 31B, 34B), and Theodosius I towards Magnus Maximus.[67] Complementing this was the well-tried strategy of emphasising, where possible, one's own connections, whether by blood or marriage, to previous emperors of good repute, whether it be Constantine's heritage from his father Constantius I and his (questionable) claim to descent from the more distant Claudius Gothicus, or the usurper Procopius reminding troops of his blood ties with the Constantinian dynasty by parading Constantius' widow and baby daughter.[68]

Given the issue of allegiances that civil war posed so acutely, a final matter for brief consideration is the fate of the troops on the losing side who survived the bloodshed of battle. Treatment varied. A substantial number of the Pompeian troops who surrendered at the battle of Thapsus (46 BC) were massacred by Caesar's soldiers, apparently against his wishes ([Caesar] *B. Afr.* 85), but this was unusual. It was more common for defeated troops to be discharged – so, e.g., the Pompeian legions at Ilerda (49 BC) and Antony's legions after Actium (31 BC) – but even more likely for them to be redeployed. 24,000 Pompeian survivors from Pharsalus were assigned to legions to protect the eastern provinces, while 14,000 of Brutus' troops from Philippi were absorbed into the legions of Antony and Octavian (perhaps to replace those of their own time-expired soldiers who were discharged).[69] Of the legions that fought for Vitellius against the Flavian forces in 69, Tacitus reports in general terms their redeployment to the Balkans where involvement in resisting a Dacian invasion aided reintegration (*Hist.* 3.35, 46), and most are found still in service later in Vespasian's reign.[70] Likewise the pattern in the civil wars of the fourth century was almost always for defeated units to be absorbed into the forces

[66] Humphries 2008.

[67] Lunn-Rockcliffe 2010 for Maximus, with further discussion of this subject in Omissi 2018.

[68] Constantine: *Pan. Lat.* 7.3, 6.2; Procopius: Amm. Marc. 26.7.10, 26.9.3. Constantine also seems to have gone so far as to question the literal legitimacy of Maxentius, thereby casting doubt on the latter's descent from the emperor Maximian (*Orig. Const.* 12).

[69] Caes. *B. Civ.* 1.85–7, App. *B. Civ.* 4.135 with Brunt 1987: 488.

[70] This applies to I Italica, XXI Rapax, XXII Primigenia, with the fate of V Alaudae uncertain. I Germanica and XV Primigenia were disbanded, though not because detachments fought for Vitellius, but because of failure to oppose the Gallic rebel Civilis.

of the victor.[71] Generally, therefore, pragmatism tended to prevail over principle. Except most obviously after his definitive victory at Actium when, for reasons of public order and cost, Octavian was understandably keen to reduce significantly the huge numbers of men under arms, the external military needs of the empire usually dictated that experienced soldiers who had not yet completed their full term be retained in service, with new oaths of loyalty presumably sworn and, sometimes, redeployment to a different part of the empire.[72]

71 Evidence detailed in Hoffmann 1969–70: vol. 2, 11 n. 41.

72 Redeployment: in addition to those noted above, some of Magnentius' units were transferred from the west to the eastern frontier by Constantius II (Amm. Marc. 18.9.3, 19.5.2), and many Isaurians were resettled on the lower Danube frontier by Anastasius (Haarer 2006: 27).

CHAPTER 5

Society and Identity

Consideration of warfare in the Roman world necessarily entails consideration of the evolving institution responsible for conducting warfare, namely the military, and its place in Roman society. This is a subject that raises important issues about military identity and how soldiers interacted with and impacted on wider society. There are a number of caveats to register at the outset. First, it is important to appreciate that 'the military' was not a static monolithic entity, but a socially diverse body of individuals whose composition changed across Roman history.[1] Secondly, the notion of 'identity' has been the subject of debate; nonetheless, although its utility has been questioned, it remains a helpful concept, provided it is recognised that identity is a fluid construct with multiple dimensions.[2] With these caveats in mind, this chapter will begin by considering how a range of features of military life served to develop a sense of military identity, of the armed forces as a distinct community, before turning to consider soldiers' interactions with wider society, with particular reference to debate as to whether the Roman military can be regarded as a 'total institution', fundamentally separate from Roman society. The third and final section focuses on a specific aspect of this larger question, namely the religious life of soldiers, and the extent to which it was distinct from and overlapped with religious practices beyond the military. (Other aspects of cultural interaction between the military and society will be considered in Chapter 6 [Section 6.2].)

5.1 Soldiers as a Community

Despite the crucial role of military power in creating the Roman empire, there is a notable thread of anti-military sentiment running through imperial literature, perhaps most obviously in Juvenal's withering critique

[1] James 2014. [2] Mattingly 2014: 38–40.

of soldiers and their legal privileges in his sixteenth satire, but also evident in Tacitus' disdainful attitude towards rank-and-file troops, and the criticisms of the high-handed behaviour of soldiers towards civilians by a variety of late Roman writers.[3] These sentiments reflect a view of soldiers as somehow separate from the rest of society and raises the question of the extent to which the military was viewed – and viewed itself – as a distinct social entity. For most of the Republican period, the question of such a distinction was largely irrelevant, since the legions traditionally comprised part-time soldiers who undertook military service during each year's campaigning season, in between periods of farming their land. This cycle was gradually eroded over the course of the Republic, as armies sometimes spent prolonged periods abroad, while during the civil wars of the first century BC troops might be under arms for extended lengths of time. However, it was the emperor Augustus' establishment of a standing army that was the most significant development in this respect. Most obviously, this formalised military service as a full-time form of employment and thereby introduced a fundamental qualitative difference to conceptions of what it meant to be a soldier. Military service may have remained a theoretical obligation for all Roman citizens, but in practice soldiering became the experience of only a minority of the citizen body, for whom (barring death on active service) it would be a continuous experience across a number of a decades of their lives and therefore central to their identity. This development was reflected in the emergence of a term for the novel phenomenon of non-combatant citizens – *pagani*.[4] It was presumably also a major reason why military epitaphs emerged as a significant category of epigraphic monument from the Augustan period onwards, whose typical locations in military camps and cemeteries imply, moreover, that their intended 'audience' was above all a military one.[5]

The sense of social separation implicit in this new regime of military service was reinforced in various ways by Augustus. At a symbolic level soldiers were assigned separate areas of seating at public entertainments in Rome and Italy (Suet. *Aug.* 44), although this will in practice have had limited practical import since another of Augustus' measures was the removal of the legions from Italy to camps on the frontiers of the empire – a much more significant step, by virtue of creating physical distance

[3] Tacitus: Carrié 1993: 105, Speidel 2012a: 178, 185. Late Roman writers: Lee 2007: 168–9, 172–3.

[4] Mohrmann 1952: 117, with references to earlier discussions, as well as consideration of how the term subsequently acquired its religious sense; Eck 2014: 11–14. The term was still used with the sense of 'civilian' in the late sixth century: Maur. *Strat.* 1.6 (4).

[5] Speidel 2014: 326.

between soldier and civilian.[6] Further Augustan measures served to consolidate the notion that soldiers were a distinctive social group. First, he exempted soldiers from the authority of their *paterfamilias* (usually their father) when it came to decisions about how they used their military pay and any other property acquired as a soldier (their *peculium castrense*).[7] While reflecting practical recognition of the difficulties of speedy communication between the frontiers and Italy, this measure nonetheless also carried considerable symbolic significance, insofar as it involved an exceptional suspension of the time-honoured principle of *patria potestas* (paternal power). It therefore accorded soldiers a distinctive privilege, which subsequent emperors added to over the course of the next century by gradually releasing soldiers from the need to adhere to the very strict legal requirements relating to the formulation of a valid will.[8] These important privileges relating to control and bequest of a soldier's property were complemented by concessions regarding law courts, where it seems that soldiers defending themselves in a legal dispute with a civilian had the right to have the case heard by a military officer in their camp. Whether this provision originated with Augustus is unclear, but it clearly gave soldiers an advantage in legal disputes, which the poet Juvenal particularly exploited for satirical ends.[9]

A second measure usually credited to Augustus by which soldiers were distinguished from civilians did not involve a legal privilege, but rather a disability – namely, the prohibition on soldiers marrying. The reason for this prohibition has been much debated,[10] but there can be no doubt that it served to differentiate soldiers from civilians in a particularly striking manner. As discussed in more detail in the next section, it did not prevent soldiers from establishing *de facto* marriages, but such relationships had no legal status and therefore entailed serious disadvantages when it came to realising the understandable desire of many soldiers to leave their property to their partner and/or any children. The emperor Hadrian made an important concession in 119 regarding the inheritance rights of children born to serving soldiers, granting them a claim on their father's property, though without the right to be designated as principal heirs.[11]

[6] This is not to deny the presence of *some* soldiers in Rome, in the form of the Praetorian Guard and other special units (Coulston 2018).

[7] Campbell 1984: 229–36.

[8] Campbell 1984: 210–29; Champlin 1991: 56–8.

[9] Campbell 1984: 254–63.

[10] Survey in Phang 2001: 325–83, who favours an ideological motivation on Augustus' part; Wells 1989 makes the case for a practical link to Augustan marriage legislation.

[11] Campbell 1984: 285–6.

The prohibition on marriage remained in place until at least the end of the second century. Septimius Severus has usually been regarded as the emperor responsible for removing the prohibition *c.*197 as part of his efforts to confirm the loyalties of the armed forces in the aftermath of the civil war which brought him to power. However, the evidence is not unambiguous (especially the crucial phrase in Herodian 3.8.5), and a recently discovered inscription implies that the prohibition was still in place in the first decade of the third century.[12] The earliest extant evidence of a legally married soldier dates from 294,[13] so the prohibition must have been lifted at some as yet unknown point during the (relatively poorly documented) third century. Whatever the precise year, however, the prohibition was in place for more than two centuries and played a formative role in shaping a distinctive military identity.

These 'outward-looking' markers which differentiated soldiers from civilians were complemented by a range of internal features of military life which served to reinforce soldiers' sense of solidarity and of belonging to a distinct community – a sentiment exploited in a speech attributed to a general by an Augustan writer which describes the military camp as 'a second homeland (*patria altera*)' and a soldier's tent as his 'home and hearth (*domus ac penates*)'.[14] The process of creating that sense of being part of a military community began with induction into military service through swearing an oath, annual repetition of which will have acted as one reminder of a soldier's obligations and identity.[15] The religious connotations of this ritual implied by the term for the oath – *sacramentum* – are corroborated by a dedication to the *genius* (guardian spirit) of the *sacramentum* by a group of veterans.[16] The soldier, then, found his military identity being defined in relation to the different organisational groups of which he became a part, moving outwards from his immediate *contubernales* (his 'tent-mates') – the fellow-soldiers with whom he shared accommodation and meals, and alongside whom he would normally expect to fight[17] – to the century or cavalry squadron (*turma*) to which his *contubernium* belonged, and then to the legion or cohort of which his century was a part (or wing (*ala*) in the case of

[12] Eck 2011; cf. Garnsey 1970. [13] Phang 2001: 100–1 (citing *Cod. Iust.* 8.46.(47).7).
[14] Livy 44.39.5; cf. Tac. *Hist.* 2.80.3 ('the camp was held dear like a home [*penates*]').
[15] Campbell 1984: 19–32. [16] *AE* 1924.135, with Speidel & Dimitrov-Milčeva 1978: 1547–8.
[17] MacMullen 1984: 443 for evidence, with Carroll 2005 for the specific subject of the role of food preparation and consumption in reinforcing communal identity, and Lendon 2006 on *contubernalis* and related terminology in military inscriptions.

cavalry squadrons).[18] Once again the deeper significance of these groupings for soldiers is suggested by surviving dedications to the *genii* of individual centuries, cohorts, legions and wings.[19] As all this implies, soldiers had a multiplicity of points of reference contributing to their military identity.

The collective identities of units were also reinforced by the use of standards as a symbolic focal point for loyalties, with those soldiers responsible for bearing them enjoying particular prestige. In addition to the well-known legionary eagles, there were also individual standards for cohorts and centuries (Veg. *Mil.* 2.13). Moreover, military standards were objects of religious reverence, with numerous allusions in the sources to their 'sacred' character.[20] The anniversary of a legion's establishment was celebrated with sacrifices as the 'birthday of the eagle', and when not on the march, a legion's standards were housed in a special shrine in the centre of the camp, (where soldiers' savings were also deposited for safekeeping).[21] The religious significance of the standards explains why their loss in battle was regarded as such a humiliation, why troops fought so fiercely to protect them from capture by the enemy, and why the emperor Augustus was able to make so much of his recovery of Roman standards from the Parthians in 20 BC.[22]

The material discussed thus far relates to the armed forces during the Principate. The upheavals of the third century resulted in significant organisational changes to the military, while Constantine's endorsement of Christianity meant that the religious climate of Late Antiquity was increasingly different, reflected (among other things) in the adoption of Christian symbols on military equipment (Figure 4). Moreover, the volume of inscriptions, which is such an important source for understanding the lives of soldiers, contracts during Late Antiquity. Nonetheless, despite the emergence of new organisational principles, unit types and military ranks, notable continuities in relation to soldiers' identities can be detected. Perhaps unsurprisingly, the military oath continued to be important, reinforced by the addition of a reference to the Trinity in the prescribed formula (Veg. *Mil.* 2.5). And again perhaps unsurprisingly, military standards continue to feature prominently in narrative sources

[18] James 1999: 18, referring to the 'nested' levels of soldierly identity. Cf. also MacMullen 1984: 446–7.
[19] Speidel & Dimitrov-Milčeva 1978: 1544–7; Stoll 2001: 311–13.
[20] Clauss 1986: 1090–92; Stoll 2001: 257–62; Töpfer 2011: 186–98; Haynes 2013: 218–20.
[21] Seston 1969: 701–2; Ando 2000: 259–69; Reddé 2004; Veg. *Mil.* 2.20 for soldiers' savings.
[22] Lee 1996: 208–9; Wheeler 2009: 33–5.

Figure 4 Fragment of relief from the Column of Theodosius I in Constantinople (late fourth century), showing troops with a shield bearing the Christian *chi-rho* symbol (the first two Greek letters of the name 'Christ').
Beyazit Museum, Istanbul/DAI Istanbul (U. Peschlow)

and military treatises, as well as inscriptions,[23] while their loss in battle remained a serious matter and their recovery a cause for celebration.[24] Constantine confirmed the religious importance of military standards with his creation of the *labarum*, bearing Christian symbols (Euseb. *Vit. Const.* 1.31), and the late sixth-century treatise attributed to the emperor Maurice includes specific provision for the blessing of the standards in the days immediately before battle.[25] Towards the end of Late Antiquity one can

[23] Fourth century: Le Bohec 2006: 88–9; Töpfer 2011: 139–40, Hebblewhite 2017: 180–84. Sixth century: Müller 1912: 125–6; Grosse 1920: 310–12; Dennis 1981. Treatises: Veg. *Mil.* 1.20, 23; 2.20; 3.5; Maur. *Strat.* 7B.16–17. Inscriptions: *signifer* (*CIL* 6.3637 [Rome, probably 312], 3.7415 [Serdica, early 4th c.]), *draconarius* (*CIL* 5.8760 [Concordia, Italy, 390s), *signiferi*, *draconarii* and *imaginiferi* (Onur 2016: 145–9, 164–6 [Perge, Pamphylia, *c.*500]). Military standards (*signa*) also feature in imperial accession ceremonies in the fifth and sixth centuries (Const. Porph. *De caer.* 1.91–4).

[24] *Proc. Bell.* 4.28.46, 7.40.42–45; Maur. *Strat.* 1.8.18.

[25] 7.1 (*Peri tou hagiazein ta banda*). Excavations of the late Roman legionary fortress at el-Lejjūn on the Arabian frontier have found evidence suggesting continuing use of the shrine for legionary standards in the later fourth century and possibly into the sixth century (Haeckel 1987: 243, 247, 254, 257–8). Intriguingly, there is also a possible literary reference to rooms for military standards in a sixth-century fort (Proc. *Aed.* 2.8.24), but 'standards' may be a metonym for military units.

even observe Christian icons starting to assume a role analogous to that of the military standard.[26]

All these focal points of identification were in some sense 'official', but there were also elements of communal identity within the armed forces which did not derive from state-instituted organisational features of the military. One category was that of 'clubs' (*collegia* or *scholae*), albeit restricted to non-commissioned officers and soldiers performing specialised tasks, as opposed to the ordinary rank-and-file. The evidence is almost entirely epigraphic and leaves room for some uncertainty about their *raison d'être* and functioning, but there are references to individual *collegia* for more than a dozen different ranks and roles, from the early second through to the mid third century. Like civilian *collegia* they appear to have functioned as part social club or 'mess', part mutual fund for retirement and burial, and sometimes had their own premises for meetings. As with so many other aspects of military life, a religious dimension is evident in the way each *collegium* had its own *genius*, whom members honoured. Many of the inscriptions are dedications for the safety of the emperor, which no doubt explains official willingness to sanction their existence. From the perspective of members, however, the most meaningful aspect is likely to have been the club embodying comradeship and a sense of 'group solidarity' with others who performed the same work: 'The members called each other *fratres*, as if they were one big family.'[27]

Informal group religious observances could play an analogous role, perhaps most strikingly exemplified by an inscription from Rome in 227 recording a dedication to Zimidrenus, the Thracian version of the deity Asclepius, by twenty-one soldiers from the Praetorian Guard who all originated from the city of Philippopolis in Thrace.[28] Even where the members of a unit had, with the passage of time, lost direct contact with the provincial location from which the unit had originated, evidence for the maintenance of cult associated with that location implies that group religious observances beyond those specified by the state could play a role in maintaining soldiers' collective identity.[29] Involvement in other 'non-official' religious cults which were not linked back to a specific location could perform a similar function, as will be considered later in this chapter (Section 5.3). However, the different points of reference for soldiers'

[26] Lee 2007: 129 for references.
[27] MacMullen 1984: 443; Shaw 1983: 144, 148 ('group solidarity'). See also Speidel & Dimitrov-Milčeva 1978: 1548–9, Campbell 1994: 136–9 (including examples of inscriptions).
[28] *ILS* 2094, with Stoll 2001: 185–6, 207. [29] Haynes 2013: 227–32.

identity outlined in this section provide an indication of some of the factors which contributed to troops viewing themselves, and being viewed, as distinct from civilian society. In concluding this section, it is worth emphasising the complexity of soldiers' military identity – or better, identities – and also noting that this is only part of the picture – that soldiers had a range of interactions with civilian society which also contributed to that identity.

5.2 Soldiers and Society

The obvious indicators of differentiation between soldiers and civilians during the Principate, from soldiers' legal privileges and disabilities (Section 5.1) to their use of distinctive slang,[30] can make it tempting to conclude that 'the legions were quintessential closed societies'.[31] That notion has been presented in a more developed form through the proposal that the Roman military be viewed as an example of a 'total institution' – a sociological model defined as 'a place of residence and work where a large number of like-situated individuals, cut off from the wider society for an appreciable period of time, together lead an enclosed, formally administered life'.[32] This model was first deployed in relation to the Roman military with specific reference to the region of Numidia, using some of the features of military life referred to in the previous section, such as the roles of religious ritual and clubs, together with arguments about the troops stationed in that region having been recruited from outside it and their tendency to form relationships with females from within the military community – these latter points in turn implying 'a terrible estrangement between the body of the soldiery and the local populace'. From this perspective, 'the Roman army was *not* halfway between the ruler and the ruled; it was the instrument of violent force wielded by the central power structure of the empire.'[33] Subsequently, evidence from the eastern fortress of Dura-Europos was presented as another possible illustration of the Roman military as a 'total institution' within the context of that community.[34]

While the attractions of this model as a heuristic device for understanding the place of the Roman military within society will be evident, its application to Dura-Europos was hampered by the limited evidence-base,

[30] Mosci Sassi 1983. [31] Horsfall 2003: 107. [32] Goffman 1961: xiii.
[33] Shaw 1983: 144, 151, reviewing Fentress 1979, with her response in Fentress 1983.
[34] Pollard 1996.

while an almost contemporaneous study of the military in a much better documented region of the empire – Egypt – presented a rather different picture, of (in the words of one reviewer) 'a kinder, gentler Roman army' which was well-integrated into wider society. Although based primarily on a range of papyrological evidence from the Principate, the fourth-century Abinnaeus archive from the Fayum, with its evidence of recruitment from the local community and regular interchange between fort and civilians, showed that this pattern continued into Late Antiquity. This view found corroboration in the admittedly more restricted data from the cemeteries of Roman Mainz where civilians were increasingly buried amongst military personnel. Subsequently the notion of the Roman army as a 'total institution' received explicit criticism with reference to the armed forces in the Near East, especially on the basis of evidence for some military religious practices involving significant civilian participation. Soldiers and religion will be considered further in the next section, but here attention will be given to other relevant aspects of military life.[35]

Roman armies and military camps were, of course, never the exclusive preserve of soldiers. Incidental references imply that, as early as the Middle Republican period, soldiers were accompanied by non-combatants (who sometimes nevertheless found themselves having, *in extremis*, to take up weapons), variously referred to as *lixae* ('camp followers') and *calones* ('soldiers' servants'), or simply *servi* ('slaves'). Given that human booty was one of the profits of successful warfare, it is unsurprising that soldiers should have had slaves, whose duties ranged from gathering firewood to foraging to cooking meals to providing sexual services. Some of these individuals were responsible for grazing horses and tending to pack animals, roles of crucial importance to the logistical functioning of Roman military forces through to Late Antiquity – which has prompted the question as to whether at least some of these individuals were owned/employed by the state.[36] Whatever the answer, these were individuals who were an integral part of military life, and to that extent might be regarded as having more limited relevance to the question of military-civilian interaction.

Other non-combatants interacting with soldiers without being integrated to the same degree included entertainers (e.g., singers, dancers,

[35] Egypt: Alston 1995 with the (not uncritical) review by Bagnall 1997, and further discussion in Alston 1999. Mainz: Hope 2001: 44. Near East: Stoll 2001 (with English summary in Stoll 2007).

[36] Soldiers' slaves: Welwei 1988: 56–112; De Ligt 2015; Speidel 1989; Roth 1999: 91–112; Thorburn 2003, with Lee 2007: 152–3 for late antique evidence. Logistical roles: Roth 1999: 102–6, with Veg. *Mil.* 3.6; *Cod. Theod.* 7.22.2.1; Maur. *Strat.* 5.1, 5.2, 7A.13, 7B.10 for Late Antiquity.

mime actors) and prostitutes (the latter not always easily distinguishable from the former). Perhaps the most notorious instance of the latter derives from the Republican period, when the protracted siege of Numantia in Spain during the 130s BC apparently resulted in a reported total of 2,000 prostitutes across the various camps surrounding the town (Livy *Per.* 57, Val. Max. 2.7.1). More detailed and mundane insight into prostitution and the Roman military is provided by documents from forts along the desert road between the Nile valley and the Red Sea in Egypt during the second century; although uncertainties remain about the interpretation of some features of these texts, a dozen letters from one location seem to show that arrangements were in the hands of private individuals, not the military, and involved one prostitute at a time being contracted out on a monthly basis to a unit, typically comprising fifteen soldiers.[37]

Another source for the siege of Numantia refers to the presence in the camps of merchants – again unsurprising, but nonetheless significant in the context of military/civilian relations. From the later Republic through the Principate, soldiers increasingly had disposable income which, along with the trend towards sedentarisation, made them an obvious market for merchants. Depending on the extent to which troops were engaged in active fighting, soldiers might also have booty to sell, including slaves, so the development of a commercial symbiosis between soldier and merchant is understandable, even if not well documented.[38] However, while these different forms of interaction are relevant to the question of military/civilian relations, they are also in a sense peripheral, since these different ways of making a living – entertainer, prostitute, merchant – were peripatetic in nature and involved transient contact. For ongoing interaction with local communities, one must turn to such issues as recruitment of troops, intermarriage and involvement in religious rituals (for this last, see Section 5.3).

With regard to recruitment the limited nature of genuinely local recruitment in north Africa has been seen as a significant feature separating the military from the communities near which they were based (recruits of north African origin seem to have come from coastal, urbanised areas distant from the military camps).[39] This pattern goes against the general assumption of increasingly localised recruitment during the Principate,

[37] Entertainers: Petrikovits 1980; Phang 2001: 244–51; Pollard 1996: 225–6. Egyptian desert: Cuvigny 2003: 383–88.
[38] See, however, Polyb. 3.82, Sall. *Iug.* 44.5, Livy 10.17.5–7.
[39] Shaw 1983: 144–7; Cherry 1998: 93–5.

although the term 'local recruitment' is a problematic one, with the word 'local' open to a variety of interpretations.[40] Nonetheless, other parts of the empire did develop more genuinely localised recruitment during this period, as with auxiliary units in the increasingly important recruiting region of Pannonia.[41] The admittedly sketchier evidence from the eastern regions of Syria and Arabia suggests that over the course of the Principate, legionary and auxiliary units stationed in these regions increasingly drew on local recruits.[42] Even if some of these were from the general region rather than the more immediate locality, the likelihood that they were Greek-speakers must have facilitated interaction.

Generalising about Late Antiquity is also difficult. On the one hand, the emphasis on the mobility of some military units, particularly in the third and fourth centuries, will have militated against troops in those units serving in the locales from which they were recruited – one has only to consider the famous career inscription of Aurelius Gaius from the reign of Diocletian, whose military service resulted in his spending time in just about every region of the empire. Moreover it sometimes made practical sense to relocate units far away from their place of origin, most obviously in the case of the Anatolian region of Isauria which became an increasingly important source of military manpower in the fifth and sixth centuries, but which also had a history of banditry and rebellion; hence it is unsurprising to find an Isaurian unit being posted from Cappadocia to Egypt in the 440s. On the other hand, when the emperor Constantius II requested that his junior emperor Julian send significant numbers of his troops from Gaul to the eastern frontier in 360, part of the reason for these troops' willingness to support Julian's usurpation was their reluctance to leave Gaul because of their local ties; admittedly many of these troops were barbarians in Roman service who had been promised they would not have to campaign beyond the Alps, but their number included troops originating in Gaul reluctant to leave their families exposed to attack from across the Rhine. Likewise, it is apparent that many soldiers in late antique Egypt were locally recruited and although some units had sometimes originated from other parts of the empire or beyond, within a few generations their numbers were clearly being replenished from local sources.[43]

[40] Haynes 2013: 123–4. [41] Haynes 2013: 128–9. [42] Pollard 2000: 114–34.

[43] Aurelius Gaius: *AE* 1981.777 (tr. Campbell 1994: 240), with Drew-Bear 1981; Wilkinson 2012. Isauria: Lee 2007: 83, with Cyr. Scyth. *V. Sabae* 1 for a *numerus Isaurorum* sent to Alexandria. Gaul: Amm. Marc. 20.4.2, 4, 10, 13 (cf. Zos. 3.3.2 for Julian recruiting locally when he first arrived in Gaul). Egypt: Bagnall 1993: 176–7; Alston 1999: 193; Keenan 1994; Bagnall and Palme 1996; Palme 2007: 253.

A related issue to that of recruitment is 'marriage' patterns (bearing in mind that troops had no right to legally recognised marriages during the Principate [cf. Section 5.1]). While it has long been appreciated that, despite the prohibition on soldiers contracting legally recognised marriages during the Principate, many Roman soldiers established quasi-marriages. However, it tended to be assumed that, firstly, their 'wives' lived in adjacent civilian settlements (the so-called *canabae* and *vici*), and second, that those women were overwhelmingly drawn from local communities. Archaeological work has increasingly highlighted evidence for the presence of females within military camps, while recent analysis of epigraphic evidence has suggested that the assumption that soldiers cohabited with local women needs modifying. The archaeological work in question has above all taken the form of analysis of small finds from within military camps, focusing on objects which are much more likely to have belonged to females and children, such as hairpins, certain types of jewellery and certain sizes of shoe, as well as the skeletons of children.[44] The epigraphic analysis has involved study of auxiliary diplomas which indicate the origin of 'wives' and which has shown that, rather surprisingly, half appear to have been women from soldiers' original homeland. This obviously will not have facilitated military interaction with the local community, as also those women from within the military community, i.e. the sisters and daughters of other soldiers. However, the latter amounted to less than one sixth of the total, while the remaining third of relationships were with local women. So while this source of 'wives' was not as significant as had usually been assumed (one of the points of the study), it remained important, at least in some parts of the empire.[45] Once again, in the north African context, it seems that the predominant pattern was for soldiers to 'marry' women from within the military community, i.e. the sisters and daughters of other soldiers, with minimal evidence of soldiers establishing relationships with local women. However, the diploma evidence indicates that care should be taken in generalising from the north African situation. And that is perhaps the most important point to highlight – the need to be aware of regional variations.[46]

The question of military marriage patterns in Late Antiquity is even more problematic. The granting of the right for soldiers to contract legal

[44] Overview of debate in Allison 2011, with detailed case studies from a sample of forts along the Rhine in Allison 2013.

[45] Greene 2015: 137–8.

[46] North Africa: Shaw 1983: 148; Cherry 1998: 101–40 (though note the critique of Cherry's onomastic methodology by Mattingly 2000: 544–5).

marriages from some point in the third century onwards was an important change from a legal perspective, but the general fall-off in the epigraphic habit during Late Antiquity (including the end of auxiliary diplomas)[47] means there is much less data on which to base analysis of the backgrounds of the women whom soldiers were marrying during these centuries, and the implications for the strength of local ties. There is certainly evidence for soldiers marrying the daughters of other soldiers,[48] but there is also, unsurprisingly, evidence for them marrying local women, perhaps best exemplified by the behaviour of troops stationed in north Africa after the imperial reconquest of the region in 533, many of whom married females from the defeated Vandal population (*Proc. Bell.* 4.14.8). However, gaining a clear sense of the relative importance of these two categories is difficult.

Beyond the specific issue of familial ties, there remains the more general question of interactions between soldiers and the broader civilian population in the course of soldiers performing their duties. On the debit side of the ledger, there is certainly plenty of evidence from all periods of Roman history of soldiers taking advantage of their official position and the latent threat of violence to benefit themselves at the expense of civilians. One of the most notorious ways in which they did this was through the requisitioning of transport, whether in the form of pack animals or carts, and the related practice of requiring householders to provide accommodation and food while en route. Of course there was official sanction for troops to commandeer transport and lodgings in certain circumstances, but as a number of well-known inscriptions from Thrace and Anatolia and papyri from Egypt during the Principate make clear, soldiers sometimes demanded that provincials provide them with resources and facilities to which they were not entitled, with the repeated attempts of emperors and/or governors to prohibit such behaviour indicating the difficulties of enforcing effective controls.[49] During the later Republican period communities in the eastern Mediterranean had already experienced periodic military impositions of this sort, often with the connivance of Roman commanders,[50] but with Augustus' establishment of a standing army which needed regular provisioning, and hence more regularised transport

[47] Salway 2014: 364–5, 381. [48] MacMullen 1963: 102–3.

[49] Mitchell 1976 includes a list of relevant sources at 111–12, while Hauken 1998 provides translations and detailed discussion of a number of inscriptions; for a more recently published edict by Hadrian: Hauken & Malay 2009. For rabbinic evidence relating to Judaea, see Isaac 1992: 291–304, and for sixth-century evidence, legal and epigraphic, see Rance 2019.

[50] Broughton 1938: 574–5.

arrangements, the opportunities for abuses are likely to have multiplied during the Principate.[51] The shift to a fiscal system whereby soldiers were paid in kind during the early centuries of Late Antiquity meant no respite from the demand for transport, while the restructuring of the armed forces so that mobile field units were based in cities and towns meant increased pressure with regard to billeting, as reflected in a papyrus letter from fourth-century Oxyrhynchus which complains about harassment and a lack of respect by billeting officers (*metatores*) who 'thrust soldiers into the houses which belong to us' without warning (*P. Oxy.* 4628). The sixth century provides a particularly graphic account of the extreme disregard which soldiers might sometimes show towards their 'hosts'. Troops billeted in Edessa in 505–6 are reported to have ejected householders from their properties, and resorted to stealing food, clothing and cattle, and raping local women: 'those who came to our assistance ostensibly as saviours . . . looted us in a manner little short of enemies' (Josh. Styl. 86 [tr. Trombley & Watt]).

Having drawn attention to the evidence for oppression of civilians by soldiers, there is an important caveat to note. Particularly during the Principate, the episodes which drew complaints often involved soldiers passing through communities, rather than troops stationed on a more permanent basis in those communities[52] – and the same qualification applies in some cases from Late Antiquity, including the example from Edessa noted above. Furthermore, there is evidence, albeit more scattered, which shows soldiers in a more positive light in their dealings with civilians. The most substantial body of relevant evidence from the Principate comprises petitions addressed to district centurions or other military officers by inhabitants of Egypt, seeking aid in cases involving such issues as assault, theft or disputes with family members or neighbours. Although the outcomes of these petitions are not known, the fact that so many such requests have survived (in excess of fifty) implies that individuals believed it was worth their while to appeal to the local military for assistance.[53] Epigraphic evidence from third-century Anatolia provides further indications of a more favourable view of soldiers. One well-known inscription records a village explicitly requesting that a soldier (*stationarius*) be posted to the village to oversee implementation of a previous official decision intended to resolve a dispute with a neighbouring village,[54] while two inscriptions record the award of honorific statues by the city of Aphrodisias

51 Mitchell 1976: 129–30. 52 Cf. Speidel 2009.
53 Alston 1995: 86–96; cf. Peachin 1999, 2007. 54 Frend 1956.

to centurions who had 'conducted [themselves] decently and bravely in the province of Asia, for [their] goodwill and affection towards [the city].'[55]

Egyptian papyri from the fourth century show military officers continuing to be a point of resort for members of local communities seeking assistance, the best known example being Abinnaeus, commander of a cavalry unit based at the Fayum fort of Dionysias in the 340s. While he undoubtedly received complaints about the behaviour of some of the soldiers under his command (*P. Abinn.* 18, 28, 48), he also received requests for help in apprehending non-military wrong-doers, akin to the petitions to centurions from earlier centuries (*P. Abinn.* 12, 15, 44, 49–50, 53, 55–6). Meanwhile an epitaph from late fourth-century Italy was arranged for a deceased non-commissioned officer by the man who had once been his host (*hospes*) and was now his heir, implying that the billeting of troops need not always end badly (*ILS* 551). There is also a certain amount of papyrological evidence from sixth-century Egypt and Palestine which shows soldiers interacting with civilians in the context of legal transactions.[56] Finally, while the behaviour of some troops at Edessa in 505–6 was appalling, the same source notes that 'there were also others who lived in an orderly fashion', while a few years earlier when the city experienced an epidemic, 'Roman soldiers established places [for treatment of the sick] . . . and took care of their expenses' (Josh. Styl. 96, 43).

Overall, then, while there is evidence from other parts of the empire and other periods of Roman history consistent with the 'estrangement between the body of the soldiery and the local populace' which has been posited for Numidia during the Principate,[57] there is also evidence for closer and more amicable ties between soldier and civilian in some communities, providing a useful warning against the temptation to over-generalise.[58] The need for recognition of the complexities of military-civilian relations is only reinforced by consideration of the religious practices of soldiers.

5.3 Soldiers and Religion

A range of religious rituals surrounded the waging of war during the Republic,[59] but most of these involved actions on the part of magistrates

[55] Roueché 1981: 113–14. For further inscriptions implying a positive view of individual soldiers, see Fuhrmann 2012: 219 n. 68, 229 n. 116.

[56] Whately (forthcoming), 14–16.

[57] Shaw 1983: 144.

[58] The New Testament provides further evidence of this mixed picture, with its references to soldiers exploiting civilians (Matt. 5.41), but also living on friendly terms with local communities (Lk. 7.1–10, Acts 10).

[59] See Section 1.1, with Rüpke 1990, Rich 2013.

or commanders, in which rank-and-file troops were not necessarily participants or even observers. Commanders did sometimes perform a ritual purification (*lustratio*) of the soldiers under their command, which presumably involved animal sacrifice, and since the occasion was sometimes followed by a speech to the troops (e.g., App. *Iber.* 19, Livy 38.12.2,), this must have been a ritual which they at least observed.[60] Soldiers also participated in triumphal ceremonies which culminated in a sacrifice to Jupiter, but in addition to the fact that by no means every general celebrated a triumph, space was limited on the Capitoline hill where this part of the ceremony took place, so soldiers' involvement must have been peripheral. The creatures (eagle, wolf, minotaur, horse, boar) that featured on military standards until Marius apparently privileged the eagle in the late second century BC (Plin. *HN* 10.16) are usually assumed to have been animal totems or palladia in origin, but whether any rituals were associated with them in the Republic is unknown. More generally, since the armed forces of the Republican period comprised Roman citizens and Italian allies, the religious traditions with which soldiers were familiar were primarily those of Italian origin, even if certain cults from other parts of the Mediterranean established a presence in the city of Rome and the Italian peninsula in the final centuries of the Republic (e.g., the cults of Cybele and of Dionysus/Bacchus).

It was the creation of a standing army by the emperor Augustus that provided both the opportunity and the impetus for religious ritual to become a more regular and visible feature in the life of soldiers: the opportunity, because the stationing of troops on a permanent basis in military camps allowed scope for the establishment of an annual cycle of state-sanctioned religious events, and the impetus, because the need to reinforce the loyalty of troops to emperors encouraged the development of a religious calendar that integrated expressions of loyalty to the emperor alongside rituals honouring traditional state deities. The primary evidence for this is the religious calendar recorded in Latin on papyrus and discovered at Dura-Europos on the Euphrates (the so-called *Feriale Duranum*). Although it derives from a fortress on the edge of the empire garrisoned by a unit of *auxilia* and dates from the 230s, it has been argued on the basis of epigraphic evidence from elsewhere in the empire that it represents the religious calendar followed in all military bases, and because the emperors who feature in it stretch all the way back to Augustus, that its third-century incarnation developed out of a calendar

[60] Wheeler 2009: 239–41.

put in place by Augustus.[61] What is striking about the calendar, with its programme of sacrifices and other rituals throughout the year, is its unswerving focus on traditional Roman state deities and festivals, alongside honours for emperors and members of the imperial family, to the exclusion of any other religious cult. Whether the degree of central initiative reflected in the calendar extended to units having official priests is less certain. Sacrificial assistants (*victimarii*) are certainly listed among the various specialised roles in the military during the Principate,[62] but their role was primarily the practical one of restraining and dispatching sacrificial animals. There are also occasional references to 'priests' in military contexts, raising the possibility of 'official' religious personnel in military units, but the evidence is sufficiently ambiguous to leave the matter uncertain, particularly given that commanders and officers were capable of overseeing the relevant rituals.[63]

Given that most military units during the Principate were stationed in locations far removed from Italy where the traditional Roman pantheon will have been much less familiar to the local population, the dimension of soldiers' religious life represented by the military calendar will have tended to reinforce any sense of the military as separate from the communities near which they were based. (This is not to overlook the fact that for the increasing numbers of soldiers of provincial origin, these deities will also have been relatively unfamiliar, at least initially (see further in Section 6.2).) In older scholarship, that sense of separation has sometimes implicitly been seen as being reinforced by soldiers' involvement in other, non-official religious cults which were thought to have particularly strong associations with the military – above all, the cult of Mithras, but also that of Jupiter Dolichenus.[64]

The cult of Mithras, by origin an Indo-Iranian deity, was associated in the Roman empire with sun worship, with the bulk of the evidence deriving from the second and third centuries. As an all-male cult with a clear hierarchical structure of 'grades' of initiation through which adherents progressed, it was long thought that Mithraism must have had particular appeal to soldiers and it was assumed that the movement of troops around the empire must have played a major part in the cult's dissemination. However, while there is undoubtedly evidence of soldiers as

61 The text is *P. Dura* 54 (Welles *et al.* 1959: 191–212, Campbell 1994: 127–30). Detailed discussion in Fink, Hoey & Snyder 1940; for the epigraphic evidence, see Fishwick 1988. See also Haynes 2013: 199–206.

62 *Dig.* 50.6.7, with Wheeler 2008: 187 n. 14 for epigraphic references.

63 Wheeler 2008.

64 Cf. Shaw 1983: 149.

participants in the cult, these larger assumptions have been shown to rest on limited evidence.[65] Merchants and travellers emerge as more plausible candidates for disseminators, while of named Mithraic dedicators recorded in inscriptions (a total well in excess of 1,000 individuals), fewer than twenty per cent were soldiers in all but a couple of the most heavily militarised provinces; the exceptions are Britain and Numidia, where there is little evidence for civilian involvement in the cult. Furthermore, there is very little evidence for the involvement of the ordinary rank-and-file in the cult: most military dedicators held some sort of rank – centurions, standard-bearers, those involved in military administration – although they were usually not senior officers. 'The cult offered such men a relatively sophisticated cosmology and soteriology, reinforced by repeated common meals, . . . which legitimated their promotion through the ranks to relative privilege and actual power.'[66] Only in Numidia and Britain, then, can a case be made for seeing Mithraism as a military cult which might have contributed to a sense of separation from civilians.

Jupiter Dolichenus was a north Syrian storm god whose cult spread westwards from Commagene during the second century. Although nothing is known of the cult's theology, nearly half of the surviving 430 dedications have a military association which prompted older studies to label it a 'military cult', until this assumption was called into question by a study that emphasised that the cult was also evidently attractive to civilians.[67] More recently, the central role of the empire-wide network of military officers in the diffusion of the cult has been emphasised, alongside the argument that the cult's civilian adherents were primarily individuals who interacted with soldiers.[68] While this view of the cult shifts the balance back towards the military, it is a more nuanced one which gives due weight to the significance of military-civilian interaction.

In addition to military and civilian involvement in the cults of Mithras and Jupier Dolichenus, soldiers in many parts of the empire participated in local cults and therefore by implication had some degree of interaction with communities adjacent to where they were stationed. One of the most sustained arguments to this effect has been made by Oliver Stoll as part of his study of the army in the eastern provinces, whose title (which might be translated 'Between Integration and Demarcation') sums up his overall view of military-civilian relations during the Principate. As part of his argument for integrative aspects of the military's behaviour, he draws on inscriptions that show soldiers making dedications to local deities such as

[65] Gordon 2009. [66] Gordon 2009: 417–18. [67] Speidel 1978. [68] Collar 2013: ch. 3.

Atargatis, Jupiter Heliopolitanus and Zeus Baitoceiceus in sanctuaries also frequented by local civilians.[69] This pattern can also be observed in Egypt, where a range of sanctuaries preserve *proskynemata* – graffiti recording requests to the local deity for protection – from both soldiers and civilians.[70] Likewise in north Africa, inscriptions from temples in the neighbourhood of the military fort at Bu Njem in Tripolitania show troops seeking the support of the local deities Jupiter Hammon, Mars Canapphar and Vanammon, perhaps in part because of local recruitment,[71] while in Numidia there is significant evidence for worship of the indigenous cult of Saturn from within the major army camp at Lambaesis.[72] Finally from the lower Rhine frontier come many dedications by soldiers to local female deities referred to as the Matronae Aufaniae.[73] The prevailing theme in these dedications is the understandable concern of soldiers to harness the protective powers of local deities, and while this may not prove definitively that soldiers interacted with local civilians in religious contexts, it shows that they had certain religious rituals in common and lends credence to the likelihood of interaction in and near sanctuaries in many parts of the empire.

Another religious context in which some soldiers and civilians might conceivably have mixed during the Principate was Christian churches, a possibility suggested by the presence of the famous third-century house church within the fortress of Dura-Europos on the Euphrates.[74] Although the house church lay outside what is thought to have been the military sector of the town and there is no explicit evidence from the house church of soldiers' presence (e.g., in the form of graffiti), there is some evidence for both soldiers and civilians frequenting other religious structures in the city.[75] In raising this possibility, a number of general points need to be registered. First, although Christians sometimes experienced persecution at the hands of the Roman authorities during the Principate, such episodes were relatively rare, were localised and were usually a case of the authorities responding to popular demand. The first empire-wide persecution initiated by an emperor did not occur until 249 in the reign of the emperor Decius,[76] with a second instalment in the late 250s under the emperor Valerian, followed by a forty-year gap until the final instalment in the early years of the fourth century – the so-called 'Great Persecution' of Diocletian. All this is to say that, apart from those relatively brief periods

69 Stoll 2001: 322–49. 70 Stoll 2009: 450–1, 453–5. 71 Brouquier-Reddé 1992: 151–9.
72 Fentress 1983: 170, *contra* Shaw 1983: 149. 73 Rüger 1987, Stoll 2007: 464–6
74 Kraeling 1967. 75 Pollard 2000: 51–2. 76 Rives 1999.

of empire-wide persecution, there is no obvious reason why individual soldiers would not have been able to frequent churches. Secondly, this presupposes that there were soldiers who were also Christians. While questions have understandably been asked as to the likelihood of Christians serving in an institution in which they were required to participate regularly in the cycle of religious rituals specified in the military calendar, there is no doubt that by the early third century at the latest there were soldiers who were also Christians; perhaps some were evangelised after enlisting, and perhaps some took the view that it was acceptable to play a passive role in official religious ceremonies.[77] By the end of the third century there was a sufficient Christian presence in the military for Diocletian to anticipate his empire-wide persecution by dismissing all Christians from the armed forces (Lactant. *De mort. pers.* 10.4), though he is unlikely to have taken such an action if their number was so significant as to jeopardise the empire's military effectiveness.

The most significant question relating to Christianity and the military, however, is how quickly the military forces were 'Christianised' once emperors who supported Christianity were in charge.[78] For within a decade of Diocletian initiating his persecution in 303, Constantine had gained power in the west and from late 312 began showing favour towards the church. Although the church itself was not a monolithic entity doctrinally, and his sons favoured different theological strands, the overall thrust of imperial religious policy across the fourth century was support for Christianity, albeit without uncompromising discrimination against traditional religious practices and cults until late in the century, and with the notable exception of Constantine's nephew Julian who tried to reinvigorate paganism during his brief reign (361–3). Julian's reign makes it clear that while troops had accepted the support of Christianity by Constantine and his sons, this did not mean enthusiastic adoption of the religion across the board, for there was no widespread military reaction to Julian reversing the trend, with only a relatively small number of troops being dismissed because of their unwillingness to participate in traditional rituals.[79]

Pressure on manpower resources arising from the losses of troops on Julian's Persian expedition, and subsequently at Adrianople in 378, will

77 Lee 2007: 177–81.

78 General discussions: Tomlin 1998; Haensch 2004; Lee 2007: 176–93. The term 'Christianisation' raises issues of definition analogous to those posed by the term 'Romanisation' (on which see further in Section 6.2), but unlike the latter, it is possible to talk in terms of 'top down' initiative, whether from emperors or the church.

79 Lenski 2002b: 274–6.

have made it difficult for his Christian successors to enforce adherence to Christianity by troops over the remaining decades of the fourth century. However, the emperor Theodosius I's general prohibition of sacrifice in the early 390s (*Cod. Theod.* 16.10.9–10, 12) will presumably have gradually had an impact on religious life in the armed forces, reinforced by the introduction of an explicitly Christian element into the military oath at some point in the later fourth century (Veg. *Mil.* 2.5). Military units had acquired their own chaplains by the 430s, if not earlier,[80] which must also have facilitated the abandonment of pagan traditions and adoption of Christian rituals by troops. Given that troops tended to be recruited from rural areas, and that rural areas tended to be more resistant to evangelisation than cities and towns, it is conceivable that over the course of the fifth century military service came to be a context in which individuals who had previously had limited exposure to Christianity became acculturated to it. By the sixth century, at any rate, it had become normal for troops to fast on the eve of Easter (Proc. *Bell.* 1.18.15–16, 37), while generals expected the display of a Christian icon to inspire the rank-and-file before going into battle (Theoph. Sim. 2.3.4–6); and the late sixth-century military treatise attributed to the emperor Maurice presupposes a framework for the life of the armed forces involving regular prayers and church services.[81] The annual cycle of rituals recorded in the *Feriale Duranum* had been well and truly displaced.[82]

80 Rance 2014a. 81 Maur. *Strat.* 2.18, 7B.17, with further detail in Lee 2007: 188–93.
82 For the issues raised by the presence in the empire's armed forces of Goths who adhered to the heterodox Christianity position of Arianism, see Greatrex 2001; Lee 2007: 192–3; Lee 2013b.

CHAPTER 6

Culture and Communication

Warfare had a range of cultural impacts in the Roman world. At the most explicitly military level, there was the interchange of military ideas, especially technological. While there was some adoption of Roman practices by other states, the balance of exchange was firmly in the other direction, with the Roman military regularly embracing effective elements of their enemies' approaches to warfare. Roman military success and imperial expansion in turn brought conquered regions into contact with elements of Roman culture, while also modifying it in the longer term. The complexities of these interactions are increasingly appreciated, but it remains possible to delineate a number of ways in which the imperial military presence had a range of cultural impacts on different regions of the empire, including the question of language and communication. This last item is also the focus of the third and final section of the chapter, which examines the use of written documentation within the military and its implications for levels of literacy in the armed forces and for the ways in which those forces exercised control and were themselves controlled.

6.1 Warfare and Cultural Interchange

On the face of it, it might seem counter-intuitive that warfare could be conducive to cultural interchange, but if so, then the history of warfare in the Roman world demonstrates otherwise. The prestige arising from Roman success in warfare encouraged other states to emulate aspects of Roman military organisation and culture, but perhaps more significant – and surprising – is the extent to which Roman military forces adopted elements of the military practices of their enemies. Yet that very tradition of openness to other military cultures was undoubtedly an important reason for Roman military success, as some Roman authors themselves recognised with the benefit of hindsight (e.g., Arr. *Tact.* 33.2). At the same time, this phenomenon challenges the utility of the very term 'Roman', with its implications of homogeneity and uniformity.

Perhaps the best known early example of emulation of Roman practices was the Seleucid monarch Antiochus IV's inclusion of 5,000 troops 'armed in the Roman manner and wearing mail shirts' in the army which he paraded at Daphne, near Antioch, in 166 BC (Polyb. 30.25.3). They constituted about 10 per cent of his forces on that occasion, and the fact that troops in mail shirts are noted as part of Seleucid forces in a battle a few years later (1 Maccab. 6.35) suggests that the unit was created for more than just display. The Roman victory over Antiochus III at Magnesia in 190 BC and the recent Roman victory over Macedonia at the battle of Pydna in 168 BC had provided clear demonstrations of the superiority of Roman troops, so it is unsurprising that such a step should have been taken, perhaps aided in this case by Antiochus having spent a short period as a political hostage in Rome. Changes to infantry organisation in the Ptolemaic army based on Roman models have also been detected in this period, albeit with little in the way of accompanying modifications to weapons.[1] The first century BC saw further developments as eastern rulers began to deploy 'Roman-style' troops against Roman armies. Mithridates of Pontus included troops equipped and trained in the Roman manner in armies which confronted Sulla and Lucullus, as too did Tigranes of Armenia when facing the latter. Deiotarus of Galatia also trained his army in the Roman style in this period, albeit not with a view to resisting Rome, and it subsequently became the basis for the *legio XXII Deiotariana*.[2] From a later period, it is thought that mail shirts were adopted in Iran in the second century as a result of Roman influence, as also the use of a reinforcing plate on helmets.[3] It seems too that contact with the empire encouraged some first-century Germanic groups to adopt elements of Roman military practice, such as the use of reserves and a clearer command structure (Tac. *Ann.* 2.45, *Germ.* 30).

The fact that mail shirts feature as a distinctive feature of Roman military dress in some of these examples of Roman influence on others is, however, somewhat ironic, since the Romans themselves are thought to have adopted this form of protection from the Celts, perhaps in the third century BC.[4] This is symptomatic of the way in which the Roman military of the Republican period owed much of its success to a willingness to learn

[1] Sekunda 2001, with Van Wees 1997. [2] Sekunda 2007: 356. [3] James 2006: 373.
[4] Robinson 1975: 164, Bishop & Coulston 2006: 63, 271. Cf. Arr. *Tact.* 43.2 for Roman use of Celtic cavalry manoeuvres.

from the strengths of different enemies encountered, as Greek commentators from this period noted. One famous example, at least as told by Polybius (1.20.9–16), was the Roman decision, early in the First Punic War, to construct a war fleet for the first time, comprising 100 quinqueremes and 20 triremes, using the model of a Carthaginian quinquereme that had been captured after running aground.[5] Another particularly important example was the Roman adoption of the so-called 'Spanish sword', which became such a feared weapon in their hands: 'The Celtiberians differ greatly from others in the construction of their swords; for it has an effective point, and can deliver a powerful downward stroke from both hands. Wherefore the Romans abandoned their ancestral swords after the [war] against Hannibal, and adopted those of the Iberians.'[6] As this implies, the Romans appear to have adopted this form of sword in the late third century, presumably as a result of their encounters with Spanish troops in Hannibal's forces. The chief advantages of the sword from the Roman viewpoint have been identified as, first, its shape – especially the longer, narrower point of its blade – which suited the fighting style of the legion better (namely, an emphasis on the use of offensive sword thrusts at close quarters with the enemy), and secondly, the blade's strength, arising from its superior method of manufacture, which meant it was 'less likely to bend or break, and perhaps retained its edge and point better.'[7]

The character of the sword used by Roman forces underwent further changes in subsequent periods, contracting in length during the first century AD, before lengthening significantly by the end of the second. These changes have been related, not to imitation of the weaponry of opponents, but rather adaptations of fighting styles to counter the strengths of those opponents. So the shift to shorter swords, alongside the adoption of curved rectangular shields in place of the long oval shields of the Republican period, has been related to the need to engage even more closely with the most dangerous enemies of that period, above all Germanic opponents, while the later move to longer slashing swords, together

[5] See Lazenby 1996: 62–4 for discussion of the historicity of Polybius' account (which on balance he favours). The quinquereme was a vessel with three banks of oars, with two rowers each on the upper two oars and one rower on the lowest, making a total of five rowers per station. However, although Polybius mentions only quinqueremes in his account of the final naval battle of the war (off the Aegates Islands) (1.59.6–8, 1.63.5), the ship timbers and bronze rams recently recovered from the site of the battle are more consistent with the smaller trireme (Tusa & Royal 2012: 41–2).

[6] *Suda* M 302 (in the *Suda On Line* translation [www.stoa.org/sol/]); this passage is thought to be an excerpt from a lost portion of Polybius' history. See also Polyb. 6.25, Diod. Sic. 23.2 and (from the Principate) Arr. *Tact.* 33.1–2.

[7] James 2011a: 79–84 (quotation at 83).

with oval shields, has been linked to the emergence of mounted enemies as a major threat.[8] However, other changes in Roman weaponry and military accoutrements in the second and third centuries have been identified as reflecting Roman adoption of features of opponents' military culture, above all aspects of central Asian steppe warfare as mediated through the Sarmatians on the lower Danube and through the Parthians and their Sasanian successors to the east. Much of this influence can be seen in the finer detail of military equipment, such as ring pommels on swords (replacing wooden handles), slide (or bridge) mounts for sword scabbards (replacing rings) and changes to metal horse harness components.[9] However, the most striking instance is the emergence of units of cavalry equipped with long two-handed lances (*contarii*, from the lance, *contus*), sometimes also heavily armoured (*cataphracti*, lit. 'defended on all sides'). Roman troops with these designations appear in the sources during the second century, as well as a cavalry unit of Sarmatians posted to Britain in the 170s, and by the fourth century they, and related units of *clibanarii*, constitute a significant element in Roman cavalry forces, with a number of arsenals specifically assigned to the manufacture of their equipment.[10] While some of these units were recruited from among Sarmatians and other ethnic groups with the requisite equestrian skills, they also appear to have drawn on manpower from elsewhere in the empire, and although some scholars have questioned the effectiveness of these Roman versions of heavy cavalry, their continued presence in Roman armies as late as the sixth century implies that they were valued.[11]

A similar pattern has been observed with respect to units of archers, whether on foot or mounted, which initially drew heavily on eastern traditions of skill with this weapon, as reflected in Parthian prowess, but also that of other eastern ethnic groups who provided units for the Roman *auxilia*. A major part of the reason for their success was their use of the composite reflex bow, which subsequently found its way into wider use in Roman forces, as recruitment of archers drew on other regions of the empire.[12] Contact with steppe peoples during Late Antiquity – the Huns from the late fourth century, and the Avars from the mid-sixth – encouraged the further evolution of Roman cavalry tactics and equipment, with particular emphasis on mounted archers, aided by adoption of heavier

[8] James 2011a: 121–5, 183–8. [9] Coulston 2003, James 2006.
[10] Eadie 1967, Coulston 2003, with Haynes 2013: 82 (for Sarmatian units).
[11] Eadie 1967 (sceptical), Speidel 1984, Rance 2007: 354–5 (more positive).
[12] Coulston 1985 (esp. 288), James 2004: 248–9, Haynes 2013: ch. 9.

composite bows and high-arched saddles from the Huns.[13] The greater prominence of mounted archers in imperial forces in this period is reflected in Procopius' famous eulogy of them in the preface of his history (*Bell.* 1.1.6–16) and their pivotal role in some of the major victories which he recounts.[14] As for the influence of the Avars, imperial cavalry had adopted various aspects of their armour and dress by the end of the sixth century, but the most significant Avar legacy was the metal stirrup (thought to have originated in the Chinese cultural sphere). Although the strengths of existing saddles meant that this did not have the dramatic impact traditionally assumed, the advent of stirrups nonetheless enhanced the capabilities of cavalry, especially when it came to archery.[15]

Developments during Late Antiquity also show that there were limits to the extent of Roman adoption of opponents' military techniques. This is especially important in relation to the empire's Germanic neighbours. The undoubted extent of imperial recruitment from Germanic groups in the fourth century, reflected in the significant number of individuals of Germanic origin who rose to senior commands in the Roman armed forces, has long prompted claims that the Roman military was 'barbarised' during Late Antiquity, with consequent loss of effectiveness. One supposed symptom of that process was the adoption of formations of Germanic origin, bearing Germanic names, such as the 'shield wall' used by late Roman infantry and referred to in one source as the *fulcum*, and an irregular cavalry formation known as the *drungus*. However, the history of these formations shows that the terms used for them in Late Antiquity do not reflect the adoption of new formations of Germanic origin, but rather the application of novel names to formations whose use by the Romans can be traced back to Principate, if not earlier. The appearance of these new terms in Late Antiquity should not therefore be taken as corroboration of the 'Germanisation' of the Roman military.[16]

How is Roman adoption of features of their opponents' military cultures to be accounted for? At the most generalised level, it can be seen as yet one more example of a common phenomenon throughout military history, whereby technology or tactics that demonstrate their effectiveness are bound to attract admirers and imitators. At the more practical level, Roman exposure to new ideas was a natural consequence of the expansion

[13] Rance 2007: 355.
[14] Rance 2005 (esp. 465–9), at the same time emphasising the continuing importance of infantry in sixth-century warfare. See also Petitjean 2014.
[15] Rance 2007: 357–8. [16] Rance 2004a, 2004b, 2014b.

of the Roman state over a large geographical area which necessarily brought it into contact with, and often control of, a wide range of groups with different military traditions. It is also worth remembering that booty included weapons and armour, offering the opportunity to study alternative technologies in detail. Yet these considerations cannot be the whole story, because it looks very much like the Roman case involved what has been described as 'an asymmetric cultural process' – that is, the Romans borrowed more heavily from neighbours than neighbours did from them. In the more specific context of interchange with Sarmatians, Parthians and Sasanians, these societies had military cultures in which equestrian skills were particularly prized and to which the Romans therefore had little to contribute. Roman society, on the other hand, had long displayed an inclusiveness that was unusual in the ancient world, reflected above all in its willingness to extend citizenship to outsiders (including freed slaves), and that inclusiveness seems to have extended to adopting advantageous features of the military practices and technologies of others. That openness has in turn been linked to Rome's foundation myths, with their emphasis on inclusion of outsiders. Whatever the explanation, however, Roman pragmatism in this area proved to be a major strength, which goes to a long way to accounting for Roman military success.[17]

6.2 The Military and Roman Culture(s)

The relationship between Roman culture and the military has traditionally been framed in terms of the role of the military in 'Romanisation', whether that be 'Romanisation' of those from non-Roman backgrounds serving in the armed forces, or the 'Romanisation' of provincial communities which had sustained contact with military units. 'Romanisation' is, however, a term that, in the context of wider discussions about the impact of Roman rule, has increasingly been viewed as problematic, for a range of reasons including 'its implications of Roman intention, agency and even of some monolithic "Roman culture."'[18] As Rome acquired control of the Italian peninsula during the middle Republican period and then the beginnings of an overseas empire, and as that overseas empire expanded, Rome's cultural

[17] The best discussions of this whole subject are those of Simon James, drawing particularly on his work on Dura-Europos and its remarkable treasure trove of military equipment: James 2004: 242–54, James 2006 (372 for 'asymmetric cultural process'). For Rome's foundation myths and their emphasis on the inclusion of outsiders, see Dench 2005.

[18] Dench 2005: 30; see also Woolf 1998: 6–7, Keay 2001: 122–4, James 2001: 194–207, Mattingly 2011: 38–41.

influence necessarily expanded, but at the same time the Roman state was absorbing, to varying degrees, conquered peoples who had their own impact on Roman identity, on what it meant to be 'a Roman' in their region of the empire. These developments were reflected to some degree in the armed forces of the Roman state, which made ever greater use of Italian manpower during the Republic and of provincial manpower during the Principate and Late Antiquity, with the result that the cultural complexion of those forces themselves increasingly became fluid with the passage of time. It is therefore unsustainable to think about the relationship of Roman culture and the military in terms of a simple model of 'the Roman army' diffusing 'Roman culture' across the regions where it was stationed. A host of complex variables, including the armed forces themselves, were present to different degrees in different parts of the empire, making it difficult to generalise about the cultural impact of the military.

This is not, however, a counsel of complete despair. For one thing, there are some features of military life and practice which maintained a degree of consistency throughout much of Roman history, most obviously the status of Latin as the pre-eminent language of the army – its 'public voice'.[19] This is not to say that units recruited from communities whose first language was not Latin did not use their mother tongue in military contexts, nor is it to suggest that the Latin language itself was immune to change generally or in military contexts. However, its use as the primary language of most military documentation and of command, training and discipline through to the end of Late Antiquity, and its status as a 'super-high' language of power in military contexts, means that this aspect of Roman culture continued to exert an influence within and beyond the armed forces – providing an important counter-example of 'top-down' cultural influence (even if the resulting grasp of Latin on the part of some soldiers was sometimes uncertain).[20] 'The Roman army was undoubtedly the most potent force during the Roman Empire behind the learning of Latin by speakers of Greek and vernacular languages, and behind the consequent spread of bilingualism.'[21]

[19] Adams 1999: 134.

[20] For Latin having 'super-high' status, see Adams 2003: 599–617 ('super-high' status is explained at 597–9, 754–5). For use of Latin commands in the late sixth century, see Rance 2004a: 269, Rance 2010; for military documentation in Latin in the sixth century, see, e.g., *P. Ryl.* 609 [AD 505]. For uncertain grasp of Latin: Adams 1994. For criticism of 'Romanisation' as reflecting a 'top-down' perspective, see, e.g., Mattingly 2011: 39.

[21] Adams 2003: 761.

Another important consideration is that the parameters changed through time, with the situation during the Republic comparatively less complex than later. During the Republic the armed forces were mostly recruited from within Italy, and although units were increasingly posted overseas for years at a time, this was rarely if ever for the twenty years or more that became standard from Augustus onwards. For the Republican period, then, the primary focus for a consideration of the relationship between Roman culture and the military is within the armed forces, more specifically the cultural impact of military service on the Italian allies whose communities had their own languages and cultural traditions. It is apparent that these units retained an independent identity within the Roman armed forces, reflected in comments about the bravery or cowardice of particular units by ancient commentators; for example, a unit of Paeligni from central Italy famously took the initiative in assaulting the feared Macedonian phalanx at the battle of Pydna in 168 BC (Plut. *Aem.* 20). Their independent identity was also reflected in the layout of Roman military camps in which allied units were assigned to stations on the periphery. The need for effective communication in their own language was no doubt an important practical consideration in all this, but although this might suggest that military service was in fact unlikely to have facilitated the exposure of the allied troops to Roman culture,[22] such a conclusion risks overlooking other practicalities. Even if troops in an individual allied unit communicated in their first language amongst themselves, they will also have needed at least some basic acquaintance with Latin when it came to co-ordinating with the legions, and Latin made obvious sense as the *lingua franca* for communication with allies from other parts of the peninsula.[23] Moreover, while camp layout may have grouped allied units separately from legions, it is inconceivable that different categories of soldier did not mingle informally during their leisure time, whether when eating, dealing with traders or gambling.[24]

One important consequence of the civil wars of the late Republic was the extension of Roman citizenship, partly through grants to individuals made by leading generals using their discretionary powers, but above all through its concession to the Italian allies in the 80s BC.[25] The allies'

22 As argued, e.g., by Pfeilschifter 2007.

23 Although there is some evidence for the allies using Italic languages other than Latin when they broke away in 90 BC, it does not indicate systematic rejection of Latin (Clackson 2015: 74–77); inscribed lead sling bullets from siege of Asculum (89 BC) show both sides using Latin (Haynes 2013: 31).

24 Rosenstein 2012b: 91–103.

25 Lavan 2019a: 29–30.

acquisition of citizenship and the absorption of their manpower directly into the legions will have strengthened the impetus to improve their fluency in Latin and to identify with the Roman state. The establishment of a standing army by Augustus with legionaries serving for at least twenty years in increasingly permanent bases in frontier provinces will have further strengthened these trends during the first century, as Italian soldiers were removed from regular contact with their home communities and placed in unfamiliar environments which will have encouraged group solidarity. At the same time, Augustus' formalisation of the role of indigenous auxiliary units and the shift away from recruitment of Italians into the legions over the course of the first century towards localised recruitment among provincials meant that the ethnic and cultural complexion of the armed forces changed and became more complex. Nonetheless, within this evolving and heterogeneous context there is evidence that troops throughout the empire were exposed on a regular basis to religious rituals that had their origins in the city of Rome. The primary evidence for this is the (Latin) religious calendar from Dura-Europos (the so-called *Feriale Duranum*). As previously noted (in Section 5.3), what is striking about the calendar is its focus on traditional Roman state deities and festivals, alongside honours for emperors and members of the imperial family, to the complete exclusion of any other religious cult. Although a 'top-down' initiative, this is not to suggest that the calendar was part of a programme to impose Roman religion on troops; it is better seen as a strategy for encouraging loyalty to the emperor and the Roman state. However, one side-effect of soldiers observing and participating in these rituals on a year-round basis throughout their period of service during the Principate must have been some degree of absorption of Roman cultural values.

Alongside religious rituals, military service during the Principate appears to have involved participation in the Roman social ritual of bathing, despite baths being associated with discipline-undermining luxury in the minds of some commentators (Veg. *Mil.* 1.3). With the advent of permanent stone-built military camps before the end of the first century, bath-houses became 'an expected part of a fort or fortress'.[26] Legionary fortresses typically had a substantial bath-house within the perimeter, while auxiliary forts had smaller bath-houses outside, but there is also evidence of less permanent wooden structures for bathing from earlier in the century.[27]

[26] Revell 2007: 230. Cf. Haynes 2013: 171.

[27] Nielsen (1990: 76) records thirty-nine military bath-houses, though this can hardly be complete; for a wooden bath-house from the earlier first century at Vindonissa, see Haynes 2013: 171.

The normality of military bathing is implied not just by the frequency with which archaeological evidence for bath-houses has been identified, but also by the fact that such structures, or relevant documentary evidence, have been found in locations where water must have been a precious resource, such as military outposts at the quarries on Mons Claudianus in the deserts of eastern Egypt and the fort at Bu Njem on the fringes of the Sahara in north Africa.[28] The continued practice of military bathing in the legions is of particular interest as the number of Italians serving declined over the course of the first century, replaced by recruits from the provinces, while the *auxilia* were by definition drawn from the provinces, so this does seem to be a case of military service exposing provincials to another aspect of Roman cultural values relating to cleanliness and social interaction[29] – and bath houses continued to be a feature of military installations during Late Antiquity.[30] This is not of course to say that washing was a concept alien to provincials prior to serving in the Roman armed forces, but rather that they learned the structured procedures of Roman bathing and the social expectations and etiquette that went with it, 'part of a shared knowledge of what it was to be Roman'.[31] Furthermore, those not serving in the military but living near fortresses and forts also appear to have made use of facilities, judging by the evidence for hairpins, jewellery and milk teeth recovered from the drains of military baths, implying participation in Roman practices beyond the armed forces, at least in the vicinity of military bases.[32]

Another, rather different aspect of Roman cultural values to which troops of provincial origin could be exposed was Roman dietary habits. Written sources indicate that core military rations comprised wheat, olive oil, wine, meat, vegetables, cheese and salt,[33] a picture corroborated by archaeological data from military sites, whether in the form of animal bones, pollen samples, or oil and wine amphorae. For soldiers recruited from regions adjacent to the Mediterranean, these foodstuffs will have been little different from what they had grown up with. However, for the

28 Mons Claudianus: Peacock & Maxfield 1997: 122–34. Bu Njem: *Ostraca* 5 (eight soldiers sent to the bath-house), 36 (one soldier sent to gather wood for the bath-house).

29 Cf. Haynes 2013: 171–4.

30 Gregory 1995: vol. 1, 142–3 (forts in third-century Gaul, fourth-century Egypt and Judaea); Darby 2015 (forts in early fourth-century Arabia); Parker 1991: 125–7 (fourth-century legionary fort in Arabia); Vanhoutte 2015: 70 (Saxon Shore forts of fourth century); Pringle 1981: 166, 235 (sixth-century Numidia).

31 Revell 2007: 235.

32 Caerleon: Zienkiewicz 1986: vol. 2, 17–18, 197 (forty-six hairpins), 223 (two milk teeth from 11–12-year-olds). Vindolanda: Birley 2009: 95 (hairpins, combs, beads, a child's shoe). Whitmore 2013 includes further examples from military baths.

33 Roth 1999: 24–6.

increasing numbers who entered military service, whether in the legions or the *auxilia*, from regions removed from the Mediterranean – above all from the northern provinces along the Danube and Rhine – some elements of the core rations will have been less familiar. The cooler climate in the north prevented the cultivation of olive trees, so olive oil had never been a standard component of the diet in these regions, while a tradition of beer-production and consumption prevailed over that of wine, and barley was generally the preferred form of grain over wheat.[34] Soldiers of northern origin serving in the north will presumably have been able to purchase beer and barley if they wished, but they also had a strong incentive to consume the core military rations with which they were supplied, because during the Principate there was a deduction from soldiers' pay against the cost of the core rations.[35] There will then have been a gradual adjustment of dietary preferences for some troops, and presumably also for those in associated communities. While the Roman state obviously did not have a monopoly on the Mediterranean diet, the fact that provincial troops for whom olive oil, wine and wheat were relative novelties became accustomed to these items as a result of military service must have meant that they identified them with the Roman state.

In order to supply olive oil to the major concentrations of troops along the Rhine and the Danube, and in Britain, there was no other option than to organise the import of the product from more southerly regions, as reflected in the remains of oil amphorae found at military sites in the north.[36] However, while this was a solution also deployed initially with respect to wheat and wine, production of these was not constrained by climate to the same degree that olive oil was. There is evidence that 'many local agricultural regimes moved into wheat production in the 2nd and 3rd centuries, ensuring that the need for long-distance shipment of such products became less necessary',[37] while production of wine in Gaul rose significantly during the first and second centuries.[38] These changes must, at least in part, have been a response to military demand, but at the same time they must also have had a gradual impact on the consumption habits of provincial populations.

This trend was not, however, uniform, as illustrated by the case of meat, where analysis of animal bone remains from the western provinces has

[34] Haynes 2013: 177 (barley), 180 (beer).
[35] For deductions for food from the pay of *auxilia*, see *P.Gen.Lat.* 1 (trans. in Campbell 1994: 23–5) with Speidel 1973.
[36] See, e.g., Funari 2002. [37] Haynes 2013: 177.
[38] MacMullen 1968: 340–1, Woolf 1998: 183.

revealed intriguing variations. Sites from Italy and southern Gaul show a marked preference for pork, sites from northern Gaul and the Rhine show an emphasis on beef, and sites in Britain favour mutton, but with a strong beef component as well. When specifically military sites in the northwest are analysed, it is apparent that the preferred meat was beef, which in turn has been interpreted as reflecting the dietary preferences of provincial troops recruited from northern Gaul and Germany. The posting of some of those troops to Britain would then account for the strong showing of beef there, prompting a leading scholar in the field to refer to the 'Gallicisation' or 'Germanisation' of meat consumption in Roman Britain.[39] There were, then, limits to Roman dietary influence.

Alongside evidence for retention of aspects of indigenous dietary habits in the armed forces can be placed other manifestations of provincial traditions in military contexts, especially in the sphere of religion and ritual. So, for example, the auxiliary unit of archers originally recruited from Syrian Emesa and posted to Intercisa on the Middle Danube continued to venerate Elagabalus, the patron deity of Emesa, while soldiers in a Tungrian cavalry unit stationed at Birrens in southern Scotland left dedications to deities from their homeland, Ricagembeda and Viradecthis.[40] Similarly, Palmyrene soldiers stationed in north Africa, Dacia and Dura-Europos made dedications to the Palmyrene deities Malakbel and Iarhibol.[41] A somewhat different type of evidence derives from a third-century site in Cumbria where various features of funerary practice – above all the inclusion of unusual iron bucket pendants among the grave goods and the cremation of horses on the funeral pyres – have suggested the presence of a cavalry unit recruited from the Middle Danube region of the empire who continued to observe indigenous burial practices.[42]

A rather different perspective on the interplay between the military and culture is provided by a well-known feature from Late Antiquity already alluded to – namely, the prominence of individuals of barbarian origin in positions of senior command. During the Principate, commanders were drawn from the social elite of empire whose careers often spanned both military and civilian duties. By the fourth century, however, a clear divide had emerged, with career soldiers from humbler social origins holding senior military posts, including a notable number of men of barbarian heritage. Some of these, such as the Frank Silvanus and the half-Vandal

[39] King 2001. [40] Haynes 2013: 227–9 (Intercisa), 231 (Scotland). [41] Dirven 1999: 181–8.
[42] Cool 2004: 464–6.

Stilicho, were second-generation immigrants and so are likely to have been better integrated into imperial society than those who had been born beyond the frontiers. Whatever individual variations there may have been, however, it remains striking that a number of these men were honoured with the award of the consulship – a position which, while it no longer entailed any power, still brought significant kudos, stemming in part from its long history and association with the Roman elite. While their educational background is unlikely to have been comparable to that of the traditional elite, some of these individuals appear to have appreciated the importance of making an effort culturally. A number can be found exchanging correspondence with members of the elite, such as the senatorial aristocrat Symmachus and the prominent teacher of rhetoric Libanius, with one (Ellebichus) requesting copies of the latter's orations and another (Stilicho) acting as patron to the poet Claudian. Of course they may well have employed well-educated secretaries to pen their letters, Ellebichus may never actually have bothered to read the orations which Libanius sent, and Stilicho's patronage of Claudian can be seen as political calculation, but these actions still show cultural awareness. Moreover, Stilicho's two daughters successively married the emperor Honorius, while the daughter of the Frank Bauto married Honorius' brother Arcadius – and the fact that her name was the Hellenised Eudoxia is itself surely significant.[43]

6.3 Written Records and Literacy in the Military[44]

In his *Epitome of Military Science*, the late Roman writer Vegetius provides an overview of the administrative arrangements underpinning the operation of Roman military forces in earlier centuries:

> The administration of the entire legion, whether matters of discipline, military duties, or finance, is written down daily in the records (*acta*) with almost greater care than fiscal or legal matters are noted down in registers. Even in peacetime, soldiers take it in turns from all centuries and *contubernia* to do daily sentry duties, whether as guards in camp or in outposts. The names of those who have done their turn are entered in lists (*brevia*) so that no-one is unfairly overburdened or given exemption. When anyone receives a leave of absence (*commeatus*), it is noted in the lists, with the number of days. (2.19 [tr. N. Milner])

[43] See further Salzman 2006, Lee 2007: 153–63, McLaughlin 2014.

[44] Helpful discussions of this subject include Bowman 1994b, Phang 2007, Haynes 2013: ch. 20; Eckardt 2018, and Speidel 2018; Stauner 2004 collates much of the evidence.

Vegetius' treatise poses many problems of interpretation and cannot automatically be treated as a simple description of the Roman armed forces in earlier centuries, but his outline of a military administrative apparatus has seemed to find ample corroboration in the significant quantities of papyri from Roman Egypt and Dura-Europos that relate in one way or another to the organisation and ordering of military activities, supported by comparable documents preserved on pottery sherds (ostraca) and wooden writing tablets from Egypt, north Africa, Germany and Britain.[45] This fascinating material provides invaluable, albeit fragmented, insights into the practical operation of the armed forces during the Principate and, to a lesser extent, Late Antiquity, unmediated by the distorting lens of Graeco-Roman literary conventions. It also has the potential to shed light on the role of the written word in the exercise of power by Roman military forces and on the cultural environment of those forces. However, the implications of this material need careful evaluation. There can be a temptation to slide imperceptibly from the presence of this military documentation into assumptions about the extent to which Roman military forces were bureaucratic institutions and the extent to which soldiers were literate. The aim of this section is to sketch out some of the implications of this material for these subjects, and its limitations.

Because the relevant papyri derive from Egypt and Dura-Europos in Mesopotamia, and because these regions came under Roman control only from the time of Augustus or later, they have the potential to shed light on the administrative character of the Roman armed forces only in the Principate and later. The same limitation applies to ostraca from Egypt and the writing tablets from Switzerland (Vindonissa) and Britain (Vindolanda), while the ostraca from north Africa (Bu Njem) date to the third century. Knowledge of the role of written records in military affairs during the Republic therefore has to rely on anecdotal comments in literary sources – and even then, most of those which provide knowledge of Republican history were written during the Principate, and so could potentially be retrojecting knowledge of practices during the Principate onto the Republican period. The Republican censors kept periodically updated records of the citizen body above all for purposes of identifying those liable for military service, and sources occasionally refer specifically to *tabulae iuniores* – records of those adult males below the age of forty-six

[45] Overview in Le Bohec 2010: 197–204. Comparable papyrus documents in Late Antiquity are rare from Egypt (Palme 2007: 259, citing examples of a military roster and documents concerning promotion and discharge), but a cache survives from Nessana in the Negev (Kraemer 1958).

who could be conscripted (e.g., Livy 24.18.7). However, these can hardly be considered military documentation comparable to the documentary sources of the Principate since they were the responsibility of civilian officials and were records which also served other, non-military purposes (e.g., tax liability). There are, however, occasional hints of other forms of record in the Republican period which can be considered military documentation and precursors of the documentary material which survives from the Principate and later. Using various forms of the word *katagraphe*, Polybius refers a number of times to records of those enlisted for particular campaigns being kept by those in command of armies (e.g., 2.24.10, 6.19.5, 9.6.6), while in his famous account of the Roman army he also states that, as part of the procedure for checking that the night watch was being properly maintained, the soldiers selected to do the checking 'report to the tribune and receive from him written orders specifying which posts they are to visit and at what time' (6.35.11). In the context of the late Republican period the historian Appian refers to documentation of soldiers' disciplinary records and to military tribunes submitting a daily register of troop numbers to the commander (*B. Civ.* 3.43, 5.46), although he might be extrapolating from his own second-century context.[46]

Although it is not surprising that such mundane procedures rarely receive mention in literary sources, it is also understandable that one of the consequences of Augustus' establishment of a standing army should have been an expansion in the range of military documentation. For example, the regularisation of discharge payments will have entailed detailed records of soldiers' years of service,[47] while the need to keep track of requests for and grants of leave will have become a more significant issue in the context of a standing army. Moreover, the increasing tendency for military units to be based for extended periods in a permanent fortress or fort must also have encouraged the development and elaboration of record-keeping. This can be seen in the variety of both posts and documentation. One second-century source differentiates between military clerks (*librarii*) responsible for granaries, those who oversaw soldiers' savings, and those who maintained records of heirless property, with a dozen other titles appearing in the sources,[48] while surviving documentary evidence ranges

[46] For caution, see Harmand 1967: 200–1; for a more positive assessment, see Pearson 2016.

[47] Cf. Bowman 1994a: 36: 'The antecedents of the military bureaucracy must surely have existed in the late Republic, but it would be reasonable to guess that the developed form evolved in the Augustan period, as a concomitant of Augustus' army reforms and the introduction of the *aerarium militare*.'

[48] *Dig.* 50.6.7; Stauner 2004: 113–52.

across lists and letters, rosters and reports, and even daybooks recording in detail the timing and content of letters and parcels sent between forts.[49] The types of documentation that survive are important in demonstrating the ways in which the written word facilitated the exercise of power in military contexts. Documents such as strength reports, duty rosters and records of leave allowed the military command to monitor the location of soldiers, and thereby exercise greater control over those within the military forces,[50] while the so-called *renuntium* reports sent in to Vindolanda by detached units at outposts illustrate the way in which written communication enabled the Roman military to 'exert such effective control over such large areas with so few troops'.[51] Even the army's facilitation of the transmission of soldiers' private communications with family, friends and comrades has been seen as having a beneficial role in the maintenance of morale.[52]

For all the surviving quantities of military documents, however, it is important not to overplay the level of bureaucratic organisation in the military. Although extant documentary evidence from the Principate implies some level of record-keeping about the careers of individual soldiers, it remains the case that no example has been found of a dossier comprising the collected documentation from a soldier's career (e.g., initial recruitment, postings, promotions, leave and disciplinary records, discharge), such as one would expect in a modern bureaucracy.[53] Moreover, uniformity was not imposed across administrative practices in military contexts – most obviously, although Latin was the predominant language in military administration, Greek was in some cases used in units stationed in the eastern half of the empire.[54] But it has also been noted that, perhaps unremarkably, there is a lack of uniformity in the detail of certain categories of document found in different parts of the empire: 'Military strength reports from different places, though broadly comparable in type and purpose, do not perhaps allow as clear and precise a typology as has sometimes been thought',[55] while the format of leave requests found at Vindolanda differs in some of its detail from those found in Egypt.[56]

Even if the Roman military was not supported by a bureaucratic apparatus comparable to modern organisations, soldiers nonetheless lived

[49] 'Postal daybooks': Remijsen 2007: 135–8, discussing *O. Krok.* 1.
[50] Bowman 1994a: 36–40, Haynes 2013: 318–19.
[51] Bowman 1994b: 119.
[52] Speidel 2018: 189.
[53] Cosme 1993 argues for the existence of such dossiers in antiquity, Phang (2007: 289–90) is sceptical.
[54] Adams 2003: 600–8.
[55] Bowman 1994b: 120.
[56] Bowman 1994a: 39–40.

in a context in which they were regularly exposed to the written word to a degree that civilian rural inhabitants would not have been – and the latter made up the overwhelming majority of the empire's population. This has understandably prompted consideration of whether Roman soldiers enjoyed higher levels of literacy than 'ordinary' inhabitants of the empire (bearing in mind the conclusion of the most detailed study of literacy in the ancient world that it is unlikely to have exceeded ten per cent of the provincial population).[57] One oft-cited item of evidence suggests definite limits, namely a set of receipts for hay by members of a cavalry unit in the later second century where about two-thirds of the eighty recipients were unable to write their own statement of receipt and had to rely on a more literate comrade to do so on their behalf.[58] However, it has also been noted that one should not generalise unduly, but should distinguish between officers and soldiers, and between legionaries and auxiliaries (the hay recipients fell in the latter category).

Officers will have needed to be literate and are likely to have been drawn from social strata where literacy was the norm. Even Cerealis, commander of the Tungrian unit of *auxilia* at Vindolanda, had a good command of written Latin despite his origins in northern Gaul and his family only having acquired Roman citizenship relatively recently.[59] Centurions commissioned from the equestrian order will already have been literate, but even those who achieved this position by rising through the ranks must have acquired at least functional literacy in the course of their military service. Some of these were sufficiently confident to try their hands at poetry, even if the results were sometimes seriously flawed.[60] It is likely that *optiones* (deputy centurions) were also literate. This is the implication of the *renuntium* reports by *optiones* from Vindolanda, written as they were in a variety of hands, as also by the relief accompanying the epitaph of an *optio* from Chester who is represented carrying writing tablets.[61] It has been suggested that those with responsibilities requiring literacy comprised about 10 per cent of the troops in a legion of the Principate.[62] As for officers in Late Antiquity, the register of (Greek) language used in the sixth-century military treatise, the *Strategikon*, indicates that it was not a literary composition, but comprised essentially documentary materials 'aimed at middle-ranking officers ... whose literacy is assumed

57 Harris 1989: 254. 58 *P. Hamb.* 39 (trans. Fink 1971, no. 76). 59 Bowman 1994b: 111.

60 Adams 1999. Other examples of poet soldiers: Speidel 2014: 320.

61 Bowman 1994b: 119–20, 122, Speidel 2012b: 5; http://vindolanda.csad.ox.ac.uk/exhibition/avitusproc.shtml.

62 Le Bohec 2010: 197.

throughout';[63] the treatise includes a number of explicit references to the transmission of orders to and by officers 'in writing'.[64]

How far did literacy extend into the ranks of ordinary soldiers? There is a range of evidence which suggests that it went further than one might otherwise expect, given the rural origins of so many recruits. Anecdotal evidence from the civil wars of the late Republican period refers to various ways in which competitors for power used written propaganda to try to win over opposition troops, whether through firing leaflets attached to arrows into the opposition camp or sending agents among enemy troops distributing pamphlets by hand.[65] In a similar vein, some accounts of the events leading to Julian's acclamation as emperor at Paris in 360 refer to anonymous leaflets being distributed among the troops urging them to resist the emperor Constantius II's orders for the transfer of some of Julian's troops away from Gaul to the east.[66] Of course such activities need not imply widespread literacy among the rank-and-file, but it does suggest an assumption that there were enough individuals able to read out the content of this material to fellow-soldiers.

Also relevant are imperial laws which stipulate that copies of that law be posted in military camps. This is no doubt understandable in the case of measures which granted privileges specifically to soldiers, such as the emperor Hadrian's ruling on the children of soldiers which was to be 'posted at Alexandria in the barracks of the winter quarters of the *legio* III Cyrenaica and of the *legio* XXII Deioteriana', and the emperor Licinius' grant of benefits to troops in the so-called Brigetio tablet of the early fourth century, where 'the text of this our indulgence shall be inscribed in tablets of bronze and shall be dedicated among the standards in each camp'.[67] But it also sometimes applied to other documents, such as the senate's decree on Gnaeus Piso from the early first century, which was to be 'set up in the winter quarters of each legion where the standards are kept' – no doubt because the affair had raised issues regarding the loyalties of troops to the imperial family.[68] The reference to these texts being placed with the military standards might perhaps suggest that they were seen as having primarily a symbolic value, and yet in the case of the Brigetio tablet, its content was of real relevance to soldiers and it was in Licinius' interest for these benefits to be made known, while in the case of the decree on Piso,

[63] Rance 2004a: 268.
[64] Maur. *Strat.* 3.6 pr., 3.5 [l. 123], 3.11, 7A.4, 7B.16 [l. 41], 7B.17 [l. 47], 12B pr., 12B.24 [l. 30].
[65] Phang 2007: 299–300; Harris (2016: 98) minimises the significance of these episodes.
[66] Amm. Marc. 20.4.10 (*libellus*), Zos. 3.9.1 (*grammatia*). [67] *BGU* I.140, *AE* 1937.232.
[68] Quotation: l. 172 = Eck *et al.* 1996: 50; loyalties: ll. 52–7, 159–65 = Eck *et al.* 1996: 42, 48

the text was a very substantial one. Placing such texts with the standards is more likely to have had the practical purpose of ensuring they were visible in a location that soldiers frequented on a regular basis. No doubt, there was once again an implicit assumption that literate soldiers would read the text to non-literate companions, and yet the display of such texts in military camps accustomed even illiterate soldiers to the written word and its importance. Similarly the exterior wall of a building in the military camp at Bu Njem in north Africa 'appears to have acted as a sort of public noticeboard: fragmentary traces of painted inscriptions in letters 5–10 mm high suggest that this served as a place where information was disseminated to all the camp's soldiers ... The texts are very fragmentary, but what is preserved suggests that the documentation dealt with matters of local interest ... rather than being official documents issued from Rome'.[69] Some of the surviving documentary evidence also contributes to this subject. In addition to hay receipts already noted above, which show that about a third of the recipients *were* able to read and write, there is other evidence such as a detailed report on an ostracon written by an ordinary cavalryman,[70] and the dozen leave requests from Vindolanda, each written in a different hand and therefore 'coming, we may assume, from soldiers in the lower ranks'.[71] To all this can be added the detail that 20 per cent of surviving inkwells and their lids (88/440) have been found in military contexts.[72]

Vegetius indicates that literacy could be a criterion in assessing the suitability of would-be recruits,[73] but might it have been possible for ordinary soldiers to acquire literacy through their military service? One text from the Principate refers to the presence in the armed forces of 'clerks who are capable of teaching' (*librarii qui docere possint*: *Dig.* 50.6.7), but there has generally been scepticism about the idea of the Roman authorities including literacy as part of military training. If a soldier wanted to learn to read and write, or improve his skills in this area, it is more likely that he will have had to learn through his own devices, perhaps from a literate tent-mate. There will, however, have been some incentive to do so, since the various military posts which required literacy granted the status of

[69] Cooley 2012a: 275. [70] Haynes 2013: 313, discussing *O. Krok.* 87. [71] Bowman 1994a: 98.
[72] Eckardt 2018: 119.
[73] *Mil.* 2.19: 'Since there are several administrative departments in the legions which require literate soldiers, it is advisable that those approving recruits should test for tall stature, physical strength and alertness in everyone indeed, but in some the knowledge of 'symbols' and expertise in calculation and reckoning is selected.' Cf. *BGU* 423.

immunis, i.e. freedom from fatigues, and increased one's chances of promotion and improved pay.[74] More generally, soldiers – and those interacting with them – will have learned the value of writing and literacy through observing its benefits in action (e.g., letters of patronage), which will have served to promote the spread of literacy.[75]

[74] Speidel 2001, Haynes 2013: 322–3, 327.

[75] Bowman 1994b: 122–5, Haynes 2013: 329–36.

CHAPTER 7

Experiences of War

'History is, or should be, a subtle combination of empathetic imagination and critical analysis.'[1] If one accepts the validity and importance of trying to gain some sense of 'what it was like to be there', however difficult that may be, then consideration of experiences of warfare in the Roman world needs no justification. Moreover, much discussion of the subject of Roman warfare inevitably focuses on war from the 'top down' perspective of the elite, so thinking about experiences of warfare is an opportunity to consider the subject from other, 'bottom up' perspectives. Those experiences will have varied greatly, depending on a range of parameters – the period of Roman history, the region of the Roman world, and the social status and role of participants. This chapter aims to exercise due care with regard to problems of generalisation, above all by distinguishing between the experiences of soldiers and of civilians, as reflected in its two sections. Within each of these sections, however, the discussion also endeavours to recognise the changing circumstances of individuals across time and place. Limitations of evidence and space make anything approaching comprehensive coverage an unrealistic goal, but the aspects and examples discussed below have been selected with a view to focusing on subjects of central importance, while providing some sense of the variety of experiences. In view of these considerations, the first section focuses on soldiers' experiences of battle, while the second examines civilian experiences of three broad types of warfare which particularly affected them – sieges, raids and protracted wars.

7.1 Approaches to Roman Battle

Although military conflict took a variety of forms in antiquity, it is the set-piece 'pitched' battle that has long been seen as its archetypal expression in

[1] Hopkins 1999: 2.

the modern popular imagination. This is perhaps a legacy of the 'decisive battles' tradition of popularising histories, but it no doubt also ultimately owes much to the influence of classical historiography, beginning with Herodotus and the role of the land battles of Marathon and Plataea in his account of the Persian wars.[2] The prominence of accounts of battles in histories from antiquity has made them an understandable focal point in the modern study of ancient warfare, but the questions asked have changed with the passage of time, as will be discussed below. Whatever the questions, however, an essential preliminary is to recognise the character of these ancient accounts of battles, above all the influence of literary convention. Although the authors of histories in antiquity asserted the importance of truth as an ideal, they were also writing literature. This not only meant trying to maintain a certain standard of stylistic felicity in their use of language, but also meeting the expectations of their audience in terms of content – and when it came to battles this meant certain stock features, *topoi*, with regard to both larger elements such as pre-battle speeches and also more specific aspects of fighting, such as clouds of arrows and soldiers slipping on blood-soaked ground. The influence of Homeric epic loomed large here, as also that of the first acknowledged historians, Herodotus and Thucydides, whom subsequent authors aspired to emulate. Battle descriptions also offered historians in antiquity valuable opportunities to impress and entertain their audience with their literary and rhetorical skills. So while there are many ancient descriptions of battles from all periods of Roman history, these descriptions were often 'mummified by convention' which could result in the inclusion of 'generic fighting filler material'. This does not mean that they are necessarily worthless as evidence for the nature of ancient battle, but it does mean that they cannot be taken at face value without further interrogation, even when written by authors with military experience (which is rarely the case).[3] As will become apparent below, there are also other types of evidence, above all material remains of weapons and armour, which can be helpful in understanding aspects of ancient battle, and there are some textual sources which are less constrained by generic conventions, one example of which will be discussed in detail at the end of this section.

[2] This is not to endorse the 'western way of war' argument of Hanson (2000), which has received justified criticism (e.g., Lynn 2003: 1–27; Sidebottom 2004: 1–15).

[3] Oakley 1997: 3–12, Whitby 2007: 57–64 for overviews of the issues, and more detailed discussion in Lendon 2017 (quotations at 41, 162); for commonplace battle motifs in, e.g., Livy, see Oakley 1997: 83–4.

One long-standing approach to Roman battle has been that of trying to reconstruct the disposition and movements of the two armies and the tactics used, with a view to explaining the eventual outcome.[4] This is an obvious and understandable approach, which has sometimes been motivated by the practical concerns of military colleges to learn lessons from the past, even though the very different technological context of antiquity might seem to promise only limited insights. Its focus on dispositions and tactics has meant that it has necessarily had a command-centred perspective, but provided there is due acknowledgement of the source problems – both the literary constraints noted above and the simple difficulty of anyone ever knowing everything of importance that happened in the confusion of battle – it remains a legitimate approach. However, while a command-centred perspective might seem the most obvious and easily justified one, its primacy has been challenged over the past forty years or so by the so-called 'face of battle' approach. Although not primarily concerned with antiquity, the publication of John Keegan's *The Face of Battle* in 1976 has gradually had an important impact on approaches to ancient battle, especially in ancient Greek warfare, but also in the Roman world.[5] Through detailed discussions of the battles of Agincourt, Waterloo and the Somme, Keegan sought to shift the focus of analyses of battles from the commander to the ordinary soldier in the front line of combat.[6] While sometimes discussed as if it represents a single approach to battle, the 'face of battle' is a phrase that has been deployed in relation to a range of aspects, albeit ones that have as their common denominator a primary concern with the experience of the ordinary soldier.

One of those aspects is the mechanics of fighting, reflected in the way Keegan devoted sections of his analysis to (in the case of Agincourt) 'archers versus infantry and cavalry', 'cavalry versus infantry' and 'infantry versus infantry'. Although some Keegan-inspired discussions have considered a range of different interactions,[7] the importance of heavy infantry in

[4] For relatively recent examples, see Burns 1973 (Adrianople, 378) and Hammond 1988: 72–6 (Cynoscephalae, 197 BC).

[5] Keegan's opening chapter did include brief discussion of Thucydides and Caesar on battle, but his claim that Thucydides' account of Mantinea showed soldiers acting as human beings, whereas Caesar's of the Sambre presented them as automatons (1976: 63–8) has been criticised by classical scholars as misconceived and oversimplistic: Wheeler 2001: 170; Campbell 2002: 167–8; Kagan 2006: 182–9; Levene 2010: 276–7.

[6] In doing so, his concern was not to minimise the importance of command (Keegan 1976: 50–51, Keegan 1987), as some have claimed (e.g., Kagan 2006: 10–11).

[7] Goldsworthy 1996: 183–247; Daly 2002. Goldsworthy (1996: 116–70) also gives consideration to the commander's perspective, while Sabin (2007: 430–33) emphasises the importance of generalship in outcomes.

Roman fighting has understandably meant that particular attention has been paid to their role.[8] In his discussion of Republican infantry combat, Philip Sabin focused on four circumstantial features – the duration of battles for an hour or more, asymmetrical casualties, the mobility of the two fighting lines, and the role of supporting ranks – to propose a model of the mechanics of infantry fighting which he summarised as 'dynamic stand-off punctuated by episodes of hand-to-hand fighting'.[9] In his view, these features mean that Roman infantry combat cannot be viewed as a shoving match analogous to one common model of Greek hoplite warfare. This could not have been sustained for an hour or more, and would not account for the asymmetrical casualties, while the important role of the supporting ranks must have dictated a more flexible arrangement at the level of smaller units.

Reactions to this synthetic modelling have varied. On the one hand, what can be understood about legionary weaponry and armour from archaeological remains and iconographical representations implies a fighting style that lends some support to the more open Sabin model of infantry combat.[10] On the other hand, it has been noted that 'if the integrity of the front ranks was broken the line was vulnerable to attack by enemy infantry and particularly by cavalry',[11] while it has also been argued that insufficient account has been taken of the role and impact of missile weapons.[12] Unsurprisingly, too, the mechanics of battle changed over time in response to the impact of new enemies, notably the advent of nomadic horse archers in Late Antiquity. While infantry remained a crucial component in armies of that period, they increasingly played a more defensive role, were deployed in more compact formations and used a wider range of missiles than the traditional javelin (*pilum*), while the offensive initiative shifted to cavalry units by the sixth century.[13]

A second aspect of battle which has received increased attention as a result of Keegan's influence – arising from his focus on 'the will to combat'[14] – is that of motivation, morale and the factors that affected the willingness of Roman soldiers to fight and die. For his battles, especially Waterloo and the Somme, Keegan was able to draw on surviving diaries and letters to gain insight into the mindset of ordinary soldiers – a luxury generally unavailable to ancient historians. An exception of sorts is

[8] In addition to Sabin 2000 (discussed next), see Goldsworthy 1996: 191–227, Sabin 1996, Zhmodikov 2000.

[9] Sabin 2000: 15. [10] Coulston 2007. [11] Gilliver 2007: 134. [12] Wheeler 2001: 174–81.

[13] Rance 2007: 348–59, 363–71. [14] Keegan 1976: 114–16, 274–84.

the account of the Persian investment of Amida in 359 by the historian Ammianus Marcellinus who was one of the soldiers trapped inside the city – but this of course relates to a siege rather than a pitched battle, and Ammianus was a junior officer, writing, moreover, many years later and in the classicising historiographical tradition.[15] Otherwise, however, discussions of Roman morale have had to adduce, on the one hand, evidence for commonsense factors such as rewards and punishments (Section 2.2), training and the role of leadership (Section 4.1), and on the other – and more controversially – comparanda from combat in more recent periods of history, which run the risk of homogeonising battle across space, time and culture, and rely heavily on assumptions about human behavioural universals.[16]

Controversy has particularly arisen over the use of conclusions concerning the combat behaviour of American soldiers in the Second World War arising from the work of the journalist and official army historian S. L. A. Marshall, especially his 1947 study *Men Against Fire*. A fundamental premise of Marshall's work was that modern western troops have been inculcated with society's ethical and cultural values to regard 'the taking of life . . . [as] prohibited and unacceptable', so that 'fear of aggression . . . is part of the normal man's emotional make-up. This is his greatest handicap when he enters combat'.[17] In Marshall's view, this accounted for the surprising conclusion from his interviews with soldiers that no more than twenty-five per cent of American troops in the Second World War typically fired their weapons with a view to trying to kill the enemy.[18] The other conclusion which has particularly attracted the interest of military historians is Marshall's view that small-group bonding was of critical importance to overcoming fear of aggression: 'When a soldier is unknown to the men who are around him he has relatively little reason to fear losing the one thing that he is likely to value more highly than life – his reputation as a man among other men . . . It is the man whose identity is well known to his fellows who has the main chance as a battle effective.'[19]

[15] Discussion in Matthews 1989: 288–94; Kagan 2006: 23–51; Lenski 2007; Kelly 2008: 59–62; Levithan 2013: 170–204. Harrowing insight into the claustrophobic combat between Persian miners and Roman counter-miners during the siege of Dura-Europos in 256 has been provided by the extensive material remains found in the 1930s and more recently discussed by Simon James (2005, 2011b, 2013: 94–100).

[16] Goldsworthy 1996: 248–82; Lee 1996; Coulston 2013; Whately 2018.

[17] Marshall 1947: 78.

[18] Marshall 1947: 50–63.

[19] Marshall 1947: 153.

Controversy generated by Marshall's conclusions has centred above all on the statistical basis for his conclusion about a fire ratio of no more than 25 per cent: it seems that he did not interview as many men as he claimed, and none of those interviewed could recall being asked by Marshall whether he had fired his weapon.[20] Although debate continues regarding the merits of his conclusions,[21] doubts about his methodology have not always been taken into account when applying Marshall's conclusions to the Roman context, as also the applicability of fire ratios to the combat of Roman legionaries.[22] More generally, one must also question whether Roman soldiers are likely to have been inhibited by fear of aggression to the same extent as modern western soldiers. After all, the ethical and cultural values of Roman troops were formed in a very different social context, uninfluenced by Judaeo-Christian strictures against killing, at least until the final centuries of Roman history. Rather, there is good reason to believe that Romans were generally acculturated to the sight of bloodshed from an early age. Public religious rituals frequently involved the slaughter of animals, often in significant numbers and with much attendant gore; similarly, public entertainments in the amphitheatre usually entailed bloodshed, human and animal, as also the punishment of slaves and criminals; and cock-fighting, with its associated blood-letting, seems to have been a common 'play' activity for children in the Roman world.[23]

This is not to say that fear and anxiety played no part in the context of Roman battle. One has only to consider in closer detail the physical damage which ancient weapons could inflict on the human body, even with the protection of armour and shield,[24] to appreciate why familiarity with bloodshed would not have eliminated apprehension about engaging in armed conflict. Moreover, accounts of battle from all periods of Roman history attest the fear and anxiety often experienced by Roman soldiers.[25] Whether this warrants the assumption that ancient troops were liable to a form of Post-Traumatic Stress Disorder (PTSD), as some have suggested, is another matter – the significantly different cultural values and

20 Spiller 1988. 21 See, e.g., Glenn 2000, Jordan 2002, Engen 2009, 2011.

22 See, e.g., Goldsworthy 1996, with the criticisms of Wheeler 2001: 173, Sidebottom 2004: 90–1.

23 Bradley 1998: 545–56. 24 James 2010.

25 Harris 2006: 307, 311 for examples from the Republic; sources from Late Antiquity provide examples of troops fleeing when their officer was wounded or killed (Amm. Marc. 16.12.37–8 [Strasbourg, 357]; Joh. Mal. 464 [Callinicum, 531]). For panic in ancient warfare, see Wheeler 1988: 172–6 (with Roman examples at 176 n. 71).

technological capabilities of the ancient world encourage scepticism.[26] However, maintenance of morale clearly remained important in the Roman context, and in this respect the emphasis of Marshall – and others – on the importance of small-group cohesion cannot be dismissed so easily.[27]

Criticism of this aspect of Marshall's conclusions by scholars of modern military history has emphasised the fact that it 'makes no allowance for high casualties, particularly over a short period of time ... [The unit] has to absorb an increased flow of replacements, many of whom do not survive long enough to become anybody's buddy and whose names and backgrounds the unit's surviving members struggle to recall'.[28] This sort of scenario occurred in German units on the eastern front in the early 1940s, and among American divisions in the weeks following D-Day in 1944, but such considerations seem much less relevant to battle in antiquity, with the heaviest casualties thought to have occurred in the rout after a battle rather than during the fighting itself.[29] Small-group cohesion in the Roman context is likely to have been encouraged at the level of both *contubernium* (groups of eight 'tentmates') and century (larger groups of perhaps eighty men).[30] Although some care is needed in the interpretation of terms of apparent camaraderie (e.g., *contubernales, commilitones*) in the epigraphic record,[31] other sources imply the importance of such bonds in the context of battle (where a *contubernium* had a tactical function as an eight-man file in unit formations[32]), whether it be the unwillingness of generals to risk battle with 'troops newly enlisted, collected from many peoples, and not yet well enough acquainted amongst themselves to feel confidence in one another' (Livy 35.3.4), or the following vignette from Caesar's campaign in Egypt after the defeat of Pompey at Pharsalus in 48 BC: 'Caesar had repeatedly expounded all this during the preceding days to his men, to make them fight with greater courage ... The same points were made to each one by his tentmate (*contubernalis*), friend or acquaintance, who urged them to live up to their own self-esteem and the good opinion entertained of them by all the others' ([Caes.] *B. Alex.* 16).

Another criticism of an emphasis on small-group cohesion has been that 'ancient authors show far more awareness of competition and rivalry between individual soldiers (and units, under the empire) than of close ties between men'.[33] However, it is noteworthy that the competition that

[26] Melchior 2011, Crowley 2015.
[27] Others: Gray 1959: 39–51, Holmes 1985: 298–309; dismissive: Wheeler 2001: 173, Lendon 2004.
[28] Strachan 2006: 212. [29] Sabin 2000: 5–6. [30] MacMullen 1984: 443, 446.
[31] Lendon 2006. [32] Wheeler 1979: 312, Nicasie 1998: 212–13.
[33] Lendon 2004: 445; cf. Lendon 2005: 255.

manifested itself through Roman soldiers seeking to distinguish themselves by individual acts of bravery can often be seen working in a manner complementary to small-group cohesion in ancient accounts of Roman battle: 'it is possible to see cohesion at work in such acts, generating the bonds that drag other soldiers – in rivalry or fear of shame, certainly – after the more aggressive leaders.'[34]

A final approach to ancient battle that warrants mention is one that emphasises the role of unpredictability. It might seem to be stating the obvious that battles are unpredictable affairs, but a more sophisticated theoretical basis for this claim has been developed, drawing on insights from the mathematical field of 'chaos theory'. Chaos theory aims to model the behaviour of complex dynamic (or 'non-linear') systems in which small variations in initial conditions can result in large differences in a later state, as in the so-called 'butterfly effect' postulated for weather systems, where very minor perturbations in one part of the world can produce a major weather event, such as a tornado, in another. From this perspective, battle can be viewed as a 'non-linear system ... where there is not a simple relationship of cause and effect, but where different effects interact with and feed back into one another in a way that is effectively unpredictable'.[35] It has been argued that a perspective analogous to this forms an important part of Livy's approach to understanding battle, insofar as he 'regularly sees battles turning on small events'.[36] More generally, this approach implies that it is difficult for historians to analyse the reasons for the outcome of a battle and also for commanders to influence that outcome, at least once a battle is underway – which in turn could be seen as lending support to Keegan's 'face of battle' approach at the expense of command-centred approaches.[37]

Valuable insight into the unpredictability of battle, as well as what the experience of battle in the Roman world could entail more generally, is provided by an ancient account that provides an important exception to the usual issues raised by literary sources, as outlined at the start of this section, and which therefore warrants more detailed consideration. The battle in question took place on 14 April 43 BC near the village of Forum Gallorum in northern Italy. The context was the ongoing fallout after the

[34] Levithan 2013: 39; cf. Ward 2016.
[35] Levene 2010: 279–80, summarising the argument of Beyerchen 1992, while also noting its anticipation with reference to ancient warfare by Culham 1989.
[36] Levene 2010: 294 (noting a similar view at times in Caesar: *B. Civ.* 1.21.1, 3.68.1, 3.72.4).
[37] For a recent attempt to re-assert the latter, see Kagan 2006, who tries to take account of Beyerchen 1992, but unsuccessfully in Levene's view (2010: 279 n. 42).

assassination of Caesar twelve months earlier, when Caesar's lieutenant Mark Antony was initially challenging the senatorial establishment, specifically by besieging one of Caesar's assassins, Decimus Brutus, in the city of Mutina in northern Italy. Early in 43, one of the consuls for the year, Hirtius, proceeded north from Rome with an army to relieve Mutina, but held off engaging Antony until the other consul, Pansa, arrived with further forces from Rome. Pansa's forces were substantial in number – four legions – but they comprised new and inexperienced recruits. Antony tried to take advantage of this handicap by attacking Pansa's army before it could link up with Hirtius', an engagement which took place near the village of Forum Gallorum south-east of Mutina along the Aemilian Way. Various literary sources record the ensuing battle, but, unusually, there has also survived a letter to Cicero written by one of Pansa's officers, Galba, on the day after the battle, reporting in 400 or so words what had happened (the letter was subsequently preserved as part of Cicero's correspondence). Now of course the fact that this report takes the form of a letter rather than a historical narrative does not in itself mean that it can automatically be read as an unvarnished and objective account of events. Apart from anything else, it is apparent that Galba is keen to present his own conduct in the best possible light and to minimise the losses on his side. Nonetheless, although not the account of an ordinary soldier, the fact that it is an eyewitness report written in the immediate aftermath of the engagement lends it considerable significance.[38]

Perhaps unsurprisingly, the key points about the experience of battle that emerge from Galba's report all relate to the problems that officers faced in the midst of the confusion and uncertainties of battle. First, there is the problem of incomplete information available to participants. When Antony unexpectedly advanced against Pansa's forces, he did so in the belief that he was facing four legions of raw recruits. What he was unaware of, however, was that during the previous night, Hirtius had sent one of his experienced legions, the *legio Martia*, plus two praetorian cohorts (i.e., soldiers who acted as bodyguards to a commander, and who were therefore by definition high calibre troops), to meet Pansa's force and guide them to his camp. As a result, Antony's forces, comprising two experienced legions (II, XXXV) and two cohorts, found themselves confronting a much stiffer challenge than they had expected. Secondly, there is the problem of exercising control over troops in the confusion of battle. When Antony's

[38] The letter: Cic. *Ad Fam.* 10.30 (= Shackleton Bailey 1977: no. 378); context and discussion: Willcock 1995: 130–5, Osgood 2006: 47–55.

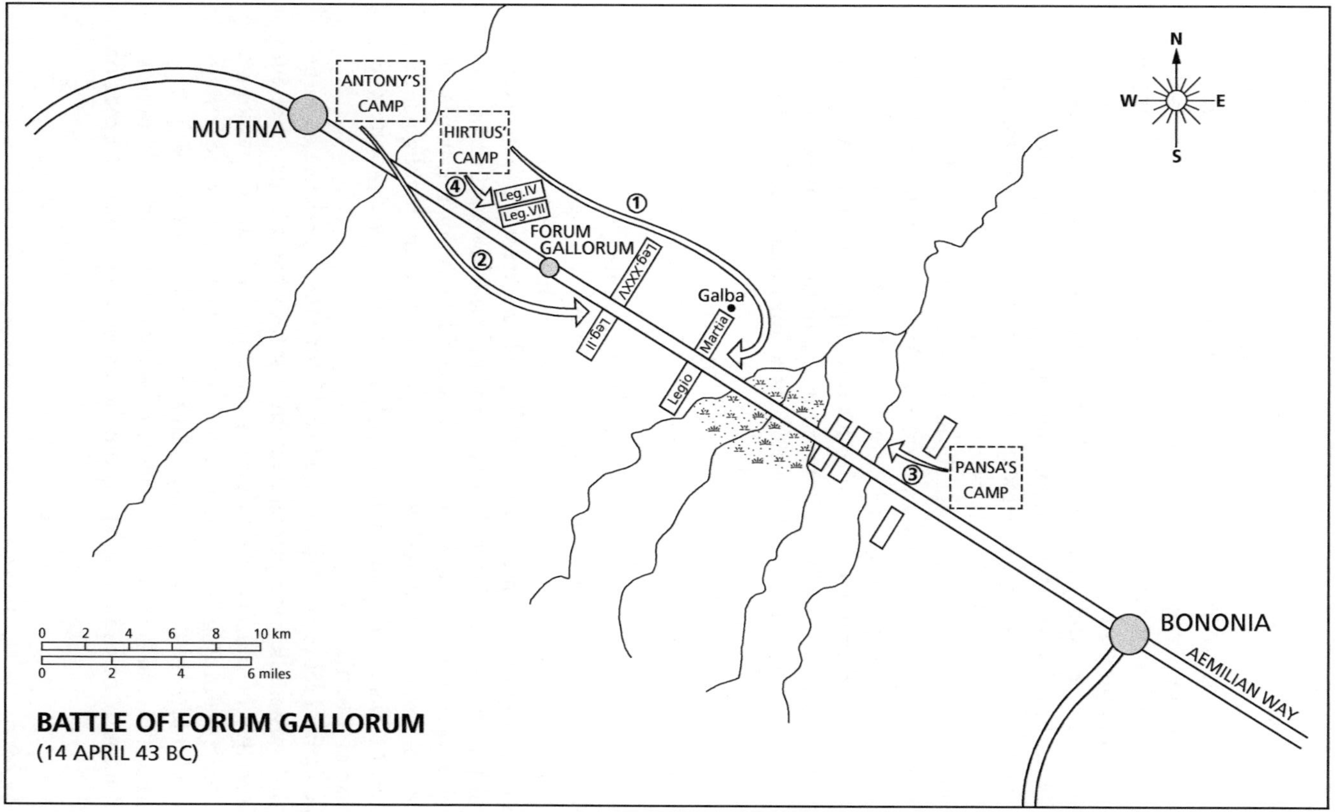

Map 6 The battle of Forum Gallorum.

forces appeared out of Forum Gallorum, the Martian legion began advancing against them on its own initiative, with Galba commenting that 'we were not able to hold them back', while the inexperienced legions were left behind and forced to 'follow them willy-nilly'. (Part of the reason for the Martian legion's eagerness to engage was their desire to avenge the deaths of some of their centurions who had been executed by Antony in Brundisium the previous year.)

The third problem that then emerged was one of co-ordination between the units within Pansa's forces; for the right wing, comprising eight cohorts of the Martian legion which Galba was helping to command, forced one of Antony's legions back to such an extent that Galba soon found himself more than half a mile ahead of the rest of Pansa's forces and in danger of being cut off and surrounded by Antony's cavalry. As Galba sought to rectify this problem, a fourth problem arose – that of distinguishing between friend and foe. In order to escape capture by Antony's cavalry, Galba rode at a gallop back towards one of the legions of recruits who, thinking he was the enemy, were about to hurl their javelins at him until at the last moment they recognised him. (This of course was a problem specific to a civil war context, which would not have arisen in conflict with non-Romans.) Antony's forces then pressed forward towards Pansa's camp, but proved unable to capture it. Even so, Antony considered himself as the victor in the engagement, until returning to Forum Gallorum his forces were unexpectedly confronted by a substantial number of Hirtius' veterans who had advanced down the Aemilian Way during the day to bring support to Pansa (Legs. IV, VII). These veterans proceeded to inflict heavy losses on Antony's forces, including the capture of two legionary eagles and sixty standards. This second phase of the battle, in which Antony suffered these heavy losses, once again highlights the difficulties of maintaining accurate knowledge of the larger picture over the course of a battle.

While Galba's letter provides some insight into the experience of battle for the ordinary soldier, it is primarily of interest for what it indicates about the perspective of officers, which perhaps accounts for its 'dispassionate tone' – 'there is nothing of the sights or sounds of the battle, of the feelings of those who fought' – and the way it 'reminds us how Romans fighting against fellow Romans could feel a sense of accomplishment'.[39] However, another account of the same battle can be used to complement that of

[39] Osgood 2006: 53.

Galba and give some sense of the experience of the rank-and-file. This is the account of the battle by the historian Appian. He was admittedly writing two centuries after the events and had literary pretensions, but he appears to have drawn on good sources (albeit perhaps more favourable to Antony),[40] judging by the accuracy of much of the circumstantial detail he provides. At one point his account focuses on the fighting between the experienced soldiers of the *legio Martia* and the legionaries on Antony's side in an area restricted by the presence of marshes alongside the Aemilian Way:

> Since the marshes and ditches gave them no chance of making outflanking movements or charging, and they were unable to push each other back, they were locked together with their swords as if in a wrestling contest. Every blow found a target, but instead of cries there were only wounds, and men dying, and groans. If a man fell, he was immediately carried away and another took his place. They had no need of encouragement or cheering on because each man's experience made him his own commanding officer. When they were tired, they separated for a few moments to recover as if they were engaged in training exercises, and then grappled with each other again. When the new recruits arrived, they were amazed to see this going on with such discipline and silence (App. *B. Civ.* 3.68 [tr. J. Carter]).

It may be that there is some rhetorical colouring here,[41] and the likelihood of orderly rest periods has been doubted,[42] but this description nonetheless conveys some sense of what it must have been like for masses of battle-hardened men to engage with swords at close quarters.

This was of course only a single battle, which makes generalisation difficult, not least because it was fought in the context of civil war. Moreover, even when fighting against non-Roman forces, the parameters will have varied to some degree depending on the identity of that enemy – and the period of Roman history, since the empire's own armed forces underwent changes and adaptations over the centuries. The role performed by different categories of soldier – whether heavy infantryman, light skirmisher, archer, cavalryman – will also have affected their experience. Despite all these caveats, however, Galba's account remains invaluable for the insights it provides into the nature and experience of battle in a central period of Roman history.

40 Shackleton Bailey 1977: vol. 2, 521. 41 Gowing 1992: 218.
42 Osgood 2006: 54, but cf. Goldsworthy 1996: 224.

7.2 Civilians and Warfare

As with soldiers, experience of warfare in the Roman world by individuals not serving in armies will have varied over time and according to geographical location. Most of the warfare of the fourth and third centuries BC relevant to Rome took place in the Italian peninsula, and so had direct consequences for many communities there in ways that became much less the case during the second century BC, as Rome's power expanded and her military campaigns were mostly conducted overseas. The so-called Social War and the civil wars of the final decades of the Republic meant that war once again impinged directly on Italy, while also continuing to affect other regions that had become imperial provinces, particularly Spain, southern Gaul and the Balkans. During the Principate warfare was mostly restricted to frontier regions, although the civil war of 68–9 impacted on northern Italy, while significant provincial revolts in the first century, notably in Judaea and southern Britain, brought substantial conflict to regions not directly adjacent to the empire's frontiers. The upheavals of the third century meant that most northern and eastern provinces experienced armed conflict with groups from outside the empire, with further significant episodes during the fourth century, while the renewed incidence of civil war sometimes affected Italy as well (notably in the early and later years of the century). Most parts of the empire, including Italy, were affected by major bouts of war during the fifth century, with the somewhat surprising exception of the eastern provinces, where Sasanian Persia's preoccupation with other problems meant a period of relative tranquillity. However, they did not escape regular conflict during the sixth and early seventh centuries, as also the Balkans, while the emperor Justinian's campaigns in the western Mediterranean meant that Italy once again became the focus of protracted fighting, along with north Africa.

Civilians usually had their closest encounters with warfare in the context of sieges, a subject for which there is substantial evidence. They would typically do so alongside a military garrison, so would not necessarily have direct involvement in fighting, but if one accepts the dictum that 'siege is the oldest form of total war,'[43] then it was bound to impact the civilian population inside a besieged city, including women and children. Siege descriptions held a revered place in the classical literary canon, ultimately traceable back to the iconic status of stories relating to Homeric

[43] Walzer 2015: 160; Levithan (2013: 20) queries the validity of the dictum for antiquity on conceptual grounds, while recognising the practical impact of siege warfare on civilians.

Figure 5 Relief from the Arch of Constantine, Rome, showing Constantine's forces besieging Verona in 312.
(W. Boucher)

Troy – Polybius (38.5) even reported Scipio Aemilianus quoting Homer as he oversaw the destruction of Carthage in 146 BC – so that, as with battles, ancient accounts of sieges can be unduly influenced by generic convention.[44] There are, however, enough detailed accounts of specific sieges by historians with direct experience of those sieges to provide 'controls', notably Julius Caesar on his siege of Alesia in Gaul in the late 50s BC, Josephus on the siege of Jerusalem in 66–70 and Ammianus on the Persian siege of Amida in 359.[45] Late Antiquity offers further valuable material in the form of Procopius' detailed narrative of the siege of Rome by the Goths in 537–8 after imperial forces had regained control of the city (*Bell.* 5.16–6.6) and the accounts of contemporary sieges in northern Mesopotamia in 502–5 preserved in the chronicle attributed to Joshua the Stylite. This latter source is particularly significant because written in Syriac, rather than Latin or Greek, and in chronicle format, both of which features made it less susceptible to the literary influence of the Graeco-Roman historiographical tradition. There is also a certain amount of visual evidence for sieges in the Roman world, which enhances understanding of this form of warfare (e.g., Figure 5).

The experience of siege warfare for civilians can usefully be differentiated into its impact during the course of the siege, on the one hand, and, on the other, the consequences for them in the event of the city being captured. During the course of a siege, non-combatant males might find themselves being drafted into guard duties on walls and gates. In Late Antiquity this sometimes involved the deployment of specific social

[44] Paul 1982. [45] Levithan 2013: 12–14.

groups, such as monks, Jews and young men from the so-called circus factions,[46] perhaps because it was assumed that common bonds would engender greater solidarity in the face of danger; during the Gothic siege of Rome in 537, Belisarius even resorted to paying civilians a daily wage for guard duty, no doubt because of the particular challenge of manning such an extensive circuit of walls and the difficulty of monitoring their performance (Proc. *Bell.* 5.25.11–12). Such civilian guards could of course become caught up in fighting during enemy assaults, as happened with the faction youths helping to defend Antioch against the Persians in 540, while some of the populace of Edessa joined imperial troops in driving off Persian forces from outside the city walls in 544 (Proc. *Bell.* 2.8.28, 2.26.7). Women and children could also sometimes be called on to take up position on the walls while soldiers made sorties, as happened at Salonae in 48 BC (Caes. *B. Civ.* 3.9), as well as helping to carry missiles, stones and even boiling oil up to defenders on the walls.[47] On occasion women were called upon to help with the manufacture of weapons and to donate their hair to the repair of damaged torsion elements in artillery catapults.[48] There was even an instance during the Hannibalic War when women from the town of Petelia in southern Italy sallied forth with the men against Carthaginian besiegers, an episode in which 'the women were no less manly than the men' (App. *Hann.* 5.29).

However, the most common ways in which a prolonged siege was likely to impact the civilian community was through food shortages and disease. Food shortages were not, of course, just an incidental consequence, but part of the strategy of besiegers to weaken the physical strength and the resolve of defenders, and to foment social discord. The besieged could respond with a range of strategies to mitigate such challenges. One obvious one was to reduce the number of mouths by expelling non-combatants, or in the case of slaves during Octavian's siege of Perusia in 41 BC, simply denying them food.[49] As food supplies dwindled, another was recourse to increasingly extreme types of *Ersatz* food, such as mice and other vermin, acorns and roots (normally used as food for farm animals), wild plants, grass and bark.[50] *In extremis*, there was sometimes even resort to

[46] Petersen 2013: 339–40. [47] Petersen 2013: 340.

[48] Weapons: Carthage, 149 BC (App. *B. Civ.* 13.93). Hair: Carthage 149 BC (Flor. 1.31.10, App. *Pun.* 13.93); Salonae, 48 BC (Caes. *B. Civ.* 3.9); Veg. *Mil.* 4.9.

[49] Expulsion: Alesia, 52 BC (Caes. *B. Gall.* 7.78); Cremna, Pisidia, 278 (Zos. 1.69); Perusia: App. *B. Civ.* 5.35. One of the sling bullets recovered from this siege bore the taunt 'You are hungry and are hiding from me' (*CIL* 1.692 [= *Eph. Epigr.* 6.84]).

[50] Garnsey 1994: 36–45, Lee 2007: 134.

cannibalism, as is reported to have happened during Sulla's siege of Athens in 87/6 BC, Theoderic Strabo's blockade of Arcadiopolis in 473, and the Roman counter-siege of Amida in 504/5.[51] Notwithstanding the influence that Thucydides' account of the Athenian plague early in the Peloponnesian War exerted on later writers, malnutrition and unsanitary conditions arising from the difficulties of disposing adequately of the dead in turn increased the vulnerability of the besieged to disease, which could spread rapidly within the confines of the city walls, as happened in Syracuse in 212 BC, Massilia in 49 BC, Amida in 359 and Rome in 409.[52] Hunger and disease also placed intense pressures on social bonds within a besieged community, with plenty of evidence of fissures opening up between soldier and civilian, and between different social groups.[53] Such pressures could sometimes spill over into violence, as in the pogrom against Jewish inhabitants of Tella in 502/3 during its siege by Persian forces (Josh. Styl. 58).

Besiegers did not always achieve their objective of capturing the city, but when they did, the consequences were invariably dire for civilian inhabitants, especially in cases where a siege was prolonged. A lengthy siege usually meant greater losses for the besiegers and therefore an understandable desire for revenge on the part of the besieging troops, whose behaviour within a captured city will, moreover, have been near impossible for a commander to control.[54] In reading ancient accounts of the denouement of sieges, the influence of traditions surrounding the fall of Homeric Troy must be borne in mind and the temptation for authors to orchestrate heightened emotions in their audience through their portrayal of the tragic fates of inhabitants.[55] At the same time, archaeological evidence leaves no doubt as to how destructive and brutal soldiers could be in their treatment of captured urban communities, with Valencia in Spain providing a particularly sobering example. Excavations in the 1990s showed the results of the capture of the city by Pompey's troops in 75 BC. Every Republican-period building was destroyed to such an extent that the earliest indications of new building work date from more than a century later. However,

51 App. *B. Mith.* 38, Malchus *fr.*2 [l.16], Josh. Styl. 77; Garnsey 1988: 28–9, Lee 2007: 134–5.

52 Livy 25.26; Caes. *B. Civ.* 2.22; Amm. Marc. 19.4, Zos. 5.39.2–3.

53 Petersen 2013: 327–36. A surprising exception is Jerusalem in 70: although factional conflict between Jewish groups had been rife beforehand, the Roman siege seems to have brought about co-operation in the face of a common enemy (Goodman 1987: 196–7).

54 Gilliver 1996: 225–6, Levithan 2013: 208–9.

55 Paul 1982; cf. also the symbolic associations and ideological resonances particularly evident in accounts of the Roman destruction of the historic cities of Carthage and Corinth in 146 BC: Purcell 1995.

it was the evidence of how some individual inhabitants were treated which particularly took excavators aback: '[They] uncovered a macabre scene consisting of several badly mutilated skeletons of some 13 males scattered over an open area, possibly the forum. The bodies display signs of torture and violent death, such as hands tied or severed, ropes around their necks, and even one individual impaled on a spear.'[56] No doubt the gruesome fates of these men were at least partly due to the political context of warfare in Spain in the period – the prolonged and bitter civil war of the 80s and 70s associated with Sertorius.[57]

Written sources from all periods of Roman history provide examples of civilian inhabitants being massacred during the sacking of a city, more often than not by Roman soldiers,[58] though with occasional instances of imperial subjects being on the receiving end, as at Gerunium in southern Italy at Carthaginian hands in 217 BC, Cremona during the civil war of 69, Amida by Persians in 503, and Milan and Tibur by Gothic forces in the mid-sixth century.[59] Distinctions were sometimes, but not always, drawn between the treatment of males of military age, on the one hand, and of women and children, on the other,[60] but for those who survived, the outcome was usually enslavement – one major aspect of the pillaging that invariably occurred. Mass enslavements in the aftermath of sieges were an integral part of Roman imperial expansion during the Republic.[61] The Principate saw a reduction in the frequency of this phenomenon, consistent with the reduced level of warfare during this period, but that frequency has perhaps been underestimated too much.[62] Josephus' history of the Jewish War of the mid-first century includes many instances of subsidiary sieges in addition to that of Jerusalem, with mass enslavements the inevitable outcome.[63]

[56] Ribera i Lacomba 2006: 80.

[57] The capture of Athens by Sulla's forces in 86 BC provides another example where archaeological evidence corroborates the destructive impact of a city's capture. While the destruction of buildings was not as thorough, those that were destroyed or badly damaged were mostly not repaired for more than half a century: Hoff 1997.

[58] Republican cases: Harris 1979: 52, 263–4, Ziolkowski 1993: 83–5. Japha, Judaea (67): Joseph. *BJ* 3.304–5. Naples (536): *Proc. Bell.* 5.10.29.

[59] Gerunium: Polyb. 3.100; Cremona: Tac. *Hist.* 3.33. Amida: Proc. *Bell.* 1.7.30. Milan (539) and Tibur (544): Proc. *Bell.* 6.21.9, 7.10.22.

[60] No distinction: Locha, Numidia (204 BC): App. *Pun.* 15; Avaricum, Gaul (52 BC): Caes. *B. Gall.* 7.28; Gamala, Judaea (67): Joseph. *BJ* 4.83. Distinction: Capsa, Numidia ((107 BC): Sall. *Iug.* 91; Volandum, Armenia (58): Tac. *Ann.* 13.39; Topeiros, Thrace (549): Proc. *Bell.* 7.38.18; Cividale (610): Paul Diac. *Hist. Lang.* 4.37.

[61] Harris 1979: 59, 63, 81; Volkmann 1990 (data organised by region and includes references from later periods).

[62] Duncan-Jones 2016: 133 provides a convenient tabulation of data.

[63] Bradley 2004: 308–9.

Mass enslavement of the inhabitants of cities captured by imperial forces continued during Late Antiquity, particularly on the empire's eastern frontier (the under-urbanised regions north of the empire offered far fewer opportunities for the capture of large communities – though this did not rule out acquisition of slaves), with the emperors Constantius II, Julian and Maurice all deporting significant numbers of prisoners from eastern communities for resettlement in imperial territory.[64] As already noted, however, inhabitants of late Roman cities within the empire were as likely to be enslaved by invaders. The Persians proved particularly keen to capture the populations of Roman cities with a view to deporting them for resettlement in Persian territory where they provided a ready supply of labour on irrigation projects and other construction schemes.[65] The best-known evidence for this derives from the mid-third century, but later kings followed suit in the fourth, sixth and early seventh century.[66] Although northern barbarians were generally much less adept at siege warfare than the Persians, the Huns and Avars proved exceptions to this general pattern. During the 440s the Huns 'sacked Ratiaria, Naissus, Philippopolis, Arcadiopolis, Constantia and very many other cities [in the Balkans] and had collected an enormous plunder and many prisoners' (Prisc. *fr*.9,4), while the Avars seized Singidunum and Anchialus in the 580s, carrying off their inhabitants into slavery (Evag. *HE* 6.10).

The fate of one of the Huns' prisoners is known in rather more detail than is usually the case for those captured after a siege, and it is a surprising story. The prisoner in question had been a merchant based in Viminacium on the lower Danube, but after the fall of the city he was assigned as a slave to one of Attila's leading men. However, after fighting for the Huns in various campaigns, his bravery and the booty he gained earned him his freedom; he married a barbarian woman with whom he had a family, enjoyed a good life in Attila's kingdom and had no desire to return to the empire with its heavy taxes and corrupt system of justice.[67] Such a happy outcome must, however, have been the exception. Much more common will have been the brutal experience evoked by an unlikely source – a didactic poem about sea life and fishing by the late second-century author Oppian. While Oppian's poetic account of the savage conflicts of marine creatures and the analogies he draws with human warfare is clearly influenced by classical literary traditions about war, he can also be seen to be

[64] Lee 2007: 137. [65] Kettenhofen 1994. [66] Lieu 1986, Morony 2004.

[67] His story is recorded by the historian Priscus who encountered him on an embassy to Attila in 449: Prisc. *fr*.11,2 [ll.407–53].

making pointed and cautionary allusions to the very real conflicts of his own day, above all the Marcomannic wars of Marcus Aurelius and Commodus, the emperors to whom his poem was dedicated. So his descriptions of human warfare, which emphasised its brutality and losses, ought not to be dismissed simply as *topoi*. At one point he describes combat between the eel and the octopus, and in depicting the eel's attempts, with its teeth, to detach the octopus from the rock to which it is clinging, the poet evokes the following image from siege warfare:

> As when a city is sacked by the hands of the enemy, and children and women are haled away as the prize of the spear, a man drags away a boy who clings to the neck and arms of his mother; the boy relaxes not his arms that are twined about her neck, nor does the wailing mother let him go, but is dragged with him herself – even so the poor body of the octopus, as he is dragged away, clings to the wet rock and does not let go.[68]

As is perhaps implicit in Oppian's vignette, another horrific but habitual accompaniment to the end of sieges was the sexual violation of women and children. Although the fate of females after the fall of Troy in Homeric epic ensured that the sexual exploitation of the defeated by the victors became a trope in classical literature, there can be no doubt that warfare has always been closely associated with sexual violence throughout history and remains so (Figure 6). It is important to appreciate, too, that 'wartime rape is not fundamentally a matter of sexual desire, but one of aggression: soldiers' rape of enemy women humiliates and emasculates enemy men, demonstrating their inability to defend their women.'[69] Roman evidence for the practice in the context of sieges takes a variety of forms, from explicit references in accounts of specific sieges, to praise of commanders for exercising restraint, to reports of besieged women opting for suicide rather than capture, to the threats of sexual violence preserved on sling bullets from the siege of Perusia in 41 BC.[70]

Another important manifestation of warfare that impacted on civilians was raiding. Because raiding often occurred on a smaller scale and by definition did not involve set-piece actions (unlike pitched battles and sieges), it features less prominently in the ancient sources. While the empire sometimes resorted to raiding as a strategy for destabilising

[68] *Halieutica* 2.313–20 (tr. A. J. Mair), with detailed treatment of the literary and contemporary resonances of Oppian in Emily Kneebone's forthcoming monograph, to whom I am grateful for discussion and advance sight of chapters from her important study.

[69] Phang 2004: 212; cf. Goldstein 2001: 363.

[70] Phang 2004: 213–21; Levithan 2013: 216–17.

Figure 6 Relief from the Column of Marcus Aurelius, Rome (late second century), showing Roman soldiers taking barbarian women captive (further discussion in Dillon 2006, who notes features that probably imply rape – drapery slipping off shoulders, and women being grabbed by the hair and wrist).
DAI Rome (F. Schlechter)

neighbouring peoples seen as threats (or used allies for this purpose),[71] it was a mode of warfare more commonly used against the empire, both because fewer of the empire's neighbours from the Principate onwards had the resources to be able to challenge the empire in set-piece military engagements and because the empire's wealth offered tempting targets for raiders. Raiding could vary significantly in scale and consequences, as is evident from two intriguing items of documentary evidence: on the one hand, a military report, preserved on a piece of pottery, of the sixty 'barbarians' who attacked one of the military stations along the Egyptian desert route from the Nile valley to the Red Sea in March 118 and carried

[71] See, e.g., Sall. BJ 54 (north Africa, late 2nd c. BC), Cassius Dio 73.2.1 (Danube, late 2nd c.), Amm. Marc. 17.10.6–7, 17.13.14, 27.10.7, 29.4.5, 30.3.1 (Rhine, mid 4th c.), Josh. Styl. 79 (Mesopotamia, early 6th c.).

off a woman and two children (*O. Krok.* 87 [ll.27–38]); on the other, an inscription on a victory altar recording the interception and defeat of a large group of Germanic Semnones and Juthungi who, after raiding into north Italy in 260, were returning northwards with 'many thousands of Italian captives' – and so must themselves have numbered in the many thousands (*AE* 1993.1231).

One way in which such raiding might impinge on civilians was when the absence of military units from a region sometimes prompted civilians to take up arms to meet the threat. This seems to have been the case in the early 170s when, with Roman forces preoccupied with the Marcomannic War, a major raid by Costobocan tribesmen across the lower Danube into the Balkans prompted a former Olympic athlete to assemble a company of men from the city of Elatea in central Greece to resist them (Paus. 10.34.5), with epigraphic evidence suggesting comparable initiatives may have been taken at Thespiae and Athens.[72] Such occasions became more common in the third century and later. The forces that intercepted the Semnonian and Juthungian raiders in 260 comprised soldiers from Raetia and the German provinces but also *populares*, 'members of the local population' (*AE* 1993.1231). Less than a decade later the Athenian historian Dexippus rallied a band of 2,000 citizens in the Attic countryside to resist Herulian raiders after they captured and sacked Athens, with the plausible suggestion that many are likely to have been youths with training through the longstanding quasi-military institution of the ephebate.[73] Similarly, in the 440s inhabitants of the Balkan city of Asemus successfully ambushed a force of Huns returning north with booty (Prisc. *fr.*9.3 ll.41–53), while the famous rearguard action that the elderly Belisarius orchestrated on the outskirts of Constantinople against a large force of Cotrigur Hun raiders in 559 involved a core of experienced troops supplemented by civilians and peasants who volunteered their services (Agath. 5.16.2–3, 5.19.6–7). A sobering counter-example is provided by the experience of Adrianople in 376, when a force of civilians who had been armed by the city's chief magistrate with a view to driving off Gothic raiders was routed, with many casualties (Amm. Marc. 31.6.2–3).

One of the most common targets for raiders was human plunder, highlighting once again the close interrelationship between warfare and slavery. In addition to the cases of the woman and children abducted in Egypt in 118 and the Italian captives being herded north by Semnones and Juthungi in 260, many further examples of this phenomenon could be

[72] Cherf 1993: 233 n. 21. [73] Fowden 1988: 53; cf. Millar 1969: 27–9.

cited, from the inhabitants of northern Anatolia carried off by seaborne Gothic raiders from across the Black Sea in the mid-third century, to nomadic Austoriani in north African Tripolitania capturing local notables and peasants from estates in the hinterland of Lepcis Magna in the early 370s, to the nearly 20,000 inhabitants of north Mesopotamian villages led away into captivity in November 502 by the Arab allies of the Persians.[74]

It is the activities of Arab raiders along the eastern fringes of the empire that provide rare detailed insight into the experience of such enslaved individuals, through the biography of a fourth-century monk, Malchus, which is all the more valuable because it 'would seem to be the only true slave narrative from antiquity'.[75] Although the account of Malchus' life by Jerome includes literary elements and has a clear religious agenda, there is good reason to think that the narrative has an authentic core, especially when assessed against other extant evidence for Arab slave-owning in this period. Malchus grew up near the important Mesopotamian centre of Nisibis, but as a young man, while travelling in the region, he was captured by camel-borne Arab raiders (*Saraceni*) and he was carried off into the desert where he was forced to work as a shepherd for his master, living on the unfamiliar pastoralist diet of milk and meat. Ostensibly as a reward for good work, but no doubt also to give him an incentive not to run away and with a view to breeding more slaves, his master tried to force him to cohabit with a fellow female captive, which Malchus resisted because he had made a prior vow of Christian chastity; when the master threatened violence, the couple made a show of agreeing to his demands, while in practice maintaining a chaste relationship. After some years, they managed to make a dramatic escape, eventually returning to Roman territory where Malchus pursued his original calling as a monk.[76] Insight into the experience of provincials enslaved through raiding is also provided by the very different evidence recovered from the gravel beds of the upper Rhine near Neupotz. Among the many hundreds of kilograms of Roman metalware (coins, weapons, tableware and the like), which has been interpreted as plunder being transported home in wagons by Germanic raiders in the mid-third century (and therefore of enormous interest in its own right), were also found shackles, implying the presence of captive provincials. It is unclear how the wagons and their contents ended up on the riverbed: most likely either the rafts ferrying them sank under their weight, or if the

[74] Anatolia: Mitchell 1993: vol. 1, 235–6. Austoriani: Amm. Marc. 28.6.4. Mesopotamia: Josh. Styl. 52; discussion and further examples in Lenski 2011a: 186–92.

[75] Lenski 2011a: 198. [76] Text and translation: Gray 2015; context: Lenski 2011b.

raiders attempted to drive them across, then the current proved too strong or the water too deep. Either way, the rescuing of any prisoners is unlikely to have been a priority.[77]

A third and final type of warfare that warrants attention for its impact on non-combatants is what might best be described as 'protracted warfare'. Protracted warfare, lasting more than a decade within the same broad region, was less common in antiquity because of logistical constraints, but for this very reason it tended to have more significant consequences for the rural economy of the relevant region. Well-documented examples of this phenomenon are provided by the Italian peninsula, during the Hannibalic War of the late third century BC (218–202) and during the Gothic war of the mid-sixth century (535–554). The economic impact of the Hannibalic War has occasioned considerable debate. The traditional view that the war had serious consequences for the rural economy of Italy, especially in the south, came under sustained criticism from Peter Brunt in the early 1970s, whose eminence ensured that this revisionist view then held sway for the next quarter of a century, until Tim Cornell presented a new evaluation of the evidence and arguments, which re-asserted important elements of the traditional picture. In particular, he emphasised two points: first, that although the scope for an invading army to destroy crops, vines and olive trees was limited, the regular presence of Carthaginian forces in the regions of southern Italy could have a serious impact on agricultural production by preventing farmers from sowing and/or harvesting crops (unless troops were deployed to protect the fields, as sometimes happened); and secondly, that it will have been small farmers who were most affected by the presence of Carthaginian forces: 'the loss of a single year's crop, or of a farm's capital equipment, would probably be enough to destroy the livelihood of a subsistence farmer and his family forever, even if they were lucky enough to survive the attack itself . . . Large numbers of such people fled to the towns and cities and failed to return once the danger was over.'[78] Despite this, however, regional variation needs to be taken into account, while the longer-term consequences of the war for population trends and land use may have been less serious than traditionally assumed.[79]

The war arising from imperial attempts to reconquer Italy from the Goths in the sixth century lasted longer than the Hannibalic War, but has occasioned much less discussion. After rapid imperial success during the

77 Catalogue of finds: Künzl 1993 (shackles: vol. 1, 365–78); discussion in Callu 1995, Todd 1996.
78 Brunt 1987: 269–77 (originally published in 1971); Cornell 1996 (quotation at 109).
79 See, e.g., Fentress 2005, Fronda 2010: 68 n. 58, de Ligt 2012: 262–3.

first five years which saw imperial forces advance from the south of Italy to Ravenna in the north, with the Gothic attempt to recapture Rome in 537–8 the only serious delay, changes of military leadership on both sides in the early 540s saw the Gothic cause re-invigorated and the imperial response handicapped by indecision and infighting amongst its generals, so that the war then dragged on for more than another decade, until the mid-550s (and in some locations in the very north of Italy, until 561). It was that decade or more of inconclusive warfare that had a serious impact on parts of the Italian peninsula. The historian Procopius presents a grim picture of the impact of war-induced famine on central regions of Italy, claiming that 50,000 inhabitants of Picenum died of starvation in the late 540s (*Bell.* 6.20.18–21). There is reason to be cautious about taking this report at face-value – Procopius may have exaggerated matters as part of his critique of the emperor Justinian – while the impact of the pandemic that swept across the Roman world in this period needs to be taken into account, as also the rapid resumption of warfare following the Lombard invasion of 568. Nonetheless there is a range of other evidence that corroborates the devastating impact of the Gothic war on much of Italy and its economy. The letters of Pope Pelagius I (556–61) 'complained that agricultural decline had severely reduced the income of the church and that the fields of Italy were so desolate that they could not be restored to cultivation';[80] archaeological investigation has shown the abandonment of villas in southern Italy during this period and this has been linked to the impact of war;[81] and a rare item of documentary evidence – a legal instrument, dated April 553 – confirmed the donation of substantial lands and their associated slaves to the church in Ravenna, including 'any of these [slaves] who have run away during these disturbed times [who] can be found' (*P. Ital.* 13). The opportunities that the disruption of war provided for the flight of slaves will have further weakened economic productivity, even if, ironically, it also shows that the usual pattern of war making slaves in the ancient world need not always hold true.

[80] Brown 1984: 6. [81] Wickham 2005: 205, 209, 477–8.

Epilogue

Warfare was a significant phenomenon throughout the history of the Roman world, exerting a fundamental influence on the evolution and fortunes of the Roman state over more than a millennium, and on the neighbours with whom the Roman state came into conflict. Warfare's role varied according to the relative size of the Roman state at particular points in time and the relative strengths of its neighbours, making it difficult to generalise. The complexities of historical causation also need to be acknowledged so as to avoid the temptation to make warfare the underlying explanation for everything in Roman history.[1] Even with those caveats, however, there is no denying the impact of war in Roman history. At the risk of over-simplification and over-generalisation, this epilogue attempts briefly to draw out some larger themes and observations.

For much of its history Roman war-making was (despite periodic defeats and setbacks) broadly successful in the most obvious sense of overcoming enemies and expanding the territory controlled by the Roman state. Warfare was primarily a means to acquire resources, human and material. Enslavement of defeated populations is a consistent theme from the Republic through to Late Antiquity, as also are the acquisition of plunder and the taxation of conquered provinces. However, one slow but steady unintended consequence of the conquest of less developed regions, especially in western Europe, was the growth of urban centres, and, with them, the gradual integration of those regions into a wider economic network (to which the flow of tax revenues contributed). Conquest was the starting point for this, and the establishment of permanent military bases along the northern peripheries of the empire in the early centuries AD also played a role. Of course those taxes in turn went towards funding

[1] Cf., e.g., growing recognition of the role of climatic change and disease in Roman history: Harper 2017, Scheidel 2018 (but note also the methodological reservations of Haldon *et al.* 2018, Sessa 2019).

the Roman state's largest budgetary item – its military forces. Roman exploitation of provincial manpower also included its use as part of those forces, setting in train a process whereby those serving (primarily in the *auxilia*) gained at least some knowledge of Latin and other aspects of Roman social practices, and eventually citizenship – all of which contributed to the integration of provincial societies into the empire. However, this was not a purely one-way process. The conquest of territories beyond Italy and their gradual incorporation into the Roman state in turn had an impact on what it meant to describe oneself as 'Roman', so warfare ultimately also had consequences for the meaning of Roman identity.

Warfare also played a major part in changing the character of the Roman state itself, albeit in less direct ways. In broad terms the acquisition of empire in the second century BC can be seen as intensifying rivalries among members of the Roman elite in the first century – so much more was now at stake – while the specific military campaigns of Pompey in the east in the 60s and of Caesar in Gaul in the 50s were important contributory factors to the civil war of 49–45 – out of which (via further rounds of civil war) there eventually emerged the Augustan regime in which power was concentrated in the hands of one individual. This fundamental change to Roman political life did not stem just from warfare, but warfare played an important role. Furthermore, this political change proved to have important consequences for Roman military forces, as Augustus formalised the gradual shift which had been evident over the course of the first century BC from a citizen militia to a standing army, with units posted to the peripheries of the empire. These changes arose from Augustus' political calculation about how best to consolidate his position, but they proved to have a whole range of further consequences – for state finances, for provincial society, for military identity, among other things. The turmoil of the mid-third century, to which warfare was a major contributor, had further consequences for the character of the Roman state, as emperors from military backgrounds became the norm in the later third and fourth century, and major re-configurations of bureaucracy and armed forces followed in an effort to meet the greater military challenges which the Roman state faced from east and north.[2] Moreover, a major factor in creating that third-century turmoil was the overthrow of the Parthian Arsacid regime by the better organised and more aggressively inclined

[2] In a similar way, the loss of Egypt and Syria – and their significant revenues – to Islamic forces in the early seventh century forced a major rethink of imperial military organisation which eventually gave rise to the system of so-called 'themes'.

Sasanian dynasty in the 220s – a development which was in part due to repeated Roman invasions of Mesopotamia during the second century undermining the authority of the Arsacids.

The long view from Republic to Late Antiquity which this study has adopted has helped to highlight important continuities through time. The ideology of victory was so deeply ingrained in Roman culture that it remained integral to the projection of imperial power even in the face of the numerous setbacks which the empire experienced during Late Antiquity, while the ideal of the citizen-soldier continued to be taken for granted long after the grant of universal citizenship in 212 had greatly reduced the significance of citizenship as a status marker. On the other hand, the advent of emperors and their concern to limit the opportunities for members of the elite to demonstrate their military prowess contributed to peace being presented as a more positive ideal than had been the case during the Republic, while the ideal of *virtus* began to lose its strong associations with manly courage in battle, as it became a more generic virtue associated with emperors and with service to them.

Potentially the most significant factor for change in military culture during Late Antiquity was imperial support for Christianity from the early fourth century onwards, but in practice Christianity did not have the radical impact which its core values might have been expected to have on war-making and those involved in it. Wars continued to be fought, soldiers still behaved badly towards civilians, and if diplomatic negotiation assumed greater prominence in Late Antiquity, this was more to do with necessity than any new-found belief in non-violence. Religious rituals associated with the military more often assumed a Christian patina rather than undergoing any fundamental change. One novelty, though hardly a positive one, was the use of religion as a justification for war, especially in the sixth and early seventh centuries. Perhaps the most significant way in which Christianity did have a positive impact in the context of warfare was when its charitable values came to the fore through bishops and clergy providing community leadership and maintaining morale in the context of the frequent sieges of this period, and using church resources to mitigate some of the impact of war through, for example, relieving food shortages and ransoming prisoners[3] – a development without obvious parallels in earlier periods of Roman history.

[3] See Lee 2007: 138, 140–1, 210.

Bibliographical Essay

Ancient Sources and Evidence

Although warfare and its wider ramifications feature to varying degrees in most literary genres from the Roman world, it is the major historical writers, with their generic focus on war and politics, that are the most important sources for the subject. Translations of Polybius, Caesar, Livy, Josephus, Tacitus, Appian, Ammianus Marcellinus, Procopius and others can be found in the Loeb Classical Library, with (more stylistically modern) alternatives usually available from Penguin and/or Oxford World Classics. These histories were all originally written in Latin or Greek. For the important early sixth-century chronicle written in the Semitic language Syriac and attributed to Joshua the Stylite, there is a good translation by Trombley and Watt (2000) in Liverpool's Translated Texts for Historians series. For military treatises, Campbell (2004) provides a useful compendium of translated extracts (to the fourth century), organised by subject, while Vegetius is translated by Nicholas Milner in the Liverpool series. There is a translation of the important late sixth-century military manual attributed to the emperor Maurice by Dennis (1984), though this will be superseded by the forthcoming translation and commentary by Philip Rance.

Inscriptions, papyri, ostraca and wooden writing tablets are invaluable documentary sources for aspects of Roman military history, but less easily accessible because scattered across many different collections and not always available in translation. However, a good selection of translated texts (to the early fourth century) can be found in Campbell (1994). For an excellent survey of archaeological evidence for ancient warfare, see James (2013), and for Roman weapons and military equipment, Bishop and Coulston (2006).

Modern Studies and Reference Works

For a narrative overview of Roman history (to 476) with a strong military dimension, see Mackay (2007), and for coverage of the whole of Late Antiquity in a similar way, Elton (2018). Woolf (2012) provides a more issue-based overview of Roman history whose focus on the theme of empire means that attention is paid to military developments in their wider context. For more detailed coverage, Routledge and Edinburgh University Press have multi-volume series by period through to the end of Late Antiquity, although both still lack at least one volume. The *Cambridge Ancient History* is even more detailed, with the final volume of the second edition completed in 2005.

There are a multitude of books on Roman military history, although few that do justice to Late Antiquity alongside the Republic and Principate. In the category of introductory overviews, Sidebottom (2004) is a stimulating introduction to many central issues and ranges from Homer to the Arab conquests, but its concise remit means that it is necessarily selective. Keppie (1984) and Roth (2009) offer more traditional but sound narrative introductions; however, they end in the second and fifth centuries respectively. Harris (2016) spans Roman history from the Republic through to the seventh century, and although the focus is not primarily military, its theme of Roman power means that military matters feature prominently, even if the treatment of Late Antiquity is sometimes less sure-footed than that of earlier periods. Oxford University Press has a series on *Ancient Warfare and Civilization* which includes a number of volumes on specific conflicts (e.g., Heather [2018]). James (2002) provides an important survey and critique of the development of British scholarship on Roman military history in the second half of the twentieth century. Brice & Roberts (2011) includes surveys of recent scholarship on Roman military history, including Late Antiquity.

A number of multi-author volumes on Roman or Greek and Roman warfare are available, although as with any edited volume, the quality of individual chapters can sometimes be variable, while the default chapter length of c.8,000 words can restrict in-depth treatment of topics. Sabin, Van Wees and Whitby (2007) is the most systematically organised. The second volume is divided into two broad periods – Late Republic and Early Empire, and the Later Empire (to the early seventh century) – within each of which there are substantial chapters of consistent quality on international relations, war, battle, armed forces, war and the state, and war and society. In line with its title's emphasis on the classical world, the

Roman chapters of Campbell and Tritle (2013) focus on the Republic and Principate, with only limited acknowledgement of Late Antiquity, but it presents themes alongside case studies, including attention to the military capabilities of some of the empire's enemies. Erdkamp (2007) is exclusively Roman in focus and gives more attention to Late Antiquity, striking a good balance between chronological and thematic chapters. Rich and Shipley (1993) is a collection of essays on war and society in the Roman world, rather than a handbook; it makes no claims to comprehensive coverage, but does include a number of influential contributions (notably those by Rich, Cornell and Woolf). Le Bohec (2015) is a multi-volume reference work on the Roman army which is systematic in its coverage of a wide range of topics in the different contexts of Republic, Principate and Late Antiquity and contains much of value, but is also uneven in quality. Sidebottom and Whitby (2018) is another multi-volume reference work, this time providing a usefully comprehensive catalogue of battles in antiquity, including discussion of the sources and the issues they pose.

There are a number of more detailed monographs that provide important treatments of particular aspects or periods. Lendon (2005) covers both Greek and Roman military history, arguing that the weight of tradition impeded innovation and offering many stimulating observations on more specific issues. Goldsworthy (1996) focuses on the war-time functioning of the Roman army in the late Republic and Principate from the perspectives of both commanders and troops, with important caveats about modernising assumptions. Campbell (2002) takes a thematic approach to warfare during the Principate and strikes a good balance between military aspects and their wider social context. James (2011a) delivers on the challenges he laid down in James (2002) by offering an insightful account of Roman military history from an archaeological perspective, with changes in sword technology providing the unifying thread (also notable for the quality of its line illustrations); the volume includes some treatment of Late Antiquity. Haynes (2013) focuses on the *auxilia* of the Principate, but does so in a way that provides an exemplary case-study in the integration of military history into wider Roman historical studies, giving attention to more traditional subjects such as recruitment and equipment alongside cultural aspects such as religion and literacy. Levithan (2013) provides a good treatment of siege warfare through to the fourth century, balancing discussion of broader issues with detailed case-studies. Erdkamp (1998) and Roth (1999) offer assessments of Roman logistical capabilities during the Republic and across the Republic and Principate respectively. For Late Antiquity Elton (1996) begins with analysis of the military capabilities of the empire's northern enemies, before providing a sound evaluation of Roman military forces in the fourth and early fifth centuries which, among other things, investigates recruitment patterns with a view to challenging traditional assumptions about 'barbarisation'. Lee (2007) is a social history of war in Late Antiquity covering a range of subjects such as infrastructure, the economic impact of war, military families and changing religious allegiances. For the ceremony of the triumph and its development over time, see Beard (2007), Goldbeck & Wienand (2016) and McCormick (1986).

Table of Significant Events

343–290 BC	Samnite Wars
280–275	War against Pyrrhus of Epirus in Italy
264–241	First Punic War against Carthage
218–201	Hannibalic War (Second Punic War)
216	Roman defeat by Hannibal at Cannae in southern Italy
202	Roman victory against Hannibal at Zama in north Africa
197	Roman victory against Philip V of Macedon at Cynoscephalae in central Greece
190	Roman victory against Seleucid king Antiochus III at Magnesia in Asia
168	Roman victory against Perseus of Macedon at Pydna in Thessaly
146	Roman forces sack Carthage and Corinth
133	Roman capture of Numantia in Spain
112–105	Jugurthine War in Numidia
105	Roman defeat by Germanic Cimbri at Arausio in southern Gaul
101	Roman victory over Cimbri at Vercellae in northern Italy
91–89	War between Rome and its Italian allies (the Social War)
88	Sulla marches on Rome
88–63	Wars against Mithridates of Pontus in Anatolia
66–62	Pompey's eastern campaigns
58–50	Caesar's Gallic campaigns
49–45	Civil war between Caesar and Pompey/Senate
44	Assassination of Caesar
43	Battle of Forum Gallorum in northern Italy
42	Battle of Philippi in northern Greece

31	Octavian (Augustus) defeats Antony and Cleopatra at Actium in western Greece
27	Octavian's adoption of the name Augustus is ratified by the Senate
AD 9	Ambush and massacre of three Roman legions under the command of Varus in the Teutoburg Forest in Germany
14	Mutiny of legions in Pannonia and on Rhine following death of Augustus
43	Roman invasion of Britain
61	Suppression of revolt of Boudicca in Britain
66–73	Suppression of Jewish revolt in Judaea
68–9	Death of Nero and civil war ('Year of the four emperors')
101–106	Trajan's conquest of Dacia (modern-day Romania)
115–117	Trajan's abortive invasion of Parthia
132–135	Suppression of Bar Kokhba revolt in Judaea
162–166	War with Parthia
170–180	Marcomannic Wars on the Danube
192–195	Civil war following death of Commodus
197–199	Septimius Severus' invasion of Parthia
*c.*226	Sasanian family overthrows Parthian regime in Persia
251	Death of emperor Decius in battle with Goths at Abrittus
260	Sasanian Persian king Shapur I defeats and captures emperor Valerian in battle in northern Mesopotamia
272	Aurelian defeats Zenobia of Palmyra in Syria
298	Galerius defeats the Persian king Narses in battle
312	Constantine defeats rival Maxentius outside Rome and begins supporting Christianity
324	Constantine defeats fellow-emperor Licinius and establishes new eastern capital of Constantinople at Byzantium
357	Julian defeats Alamannic confederation at Strasbourg
363	Julian invades Persia but dies of a wound during the campaign
376	Valens allows Goths to cross lower Danube and settle in Thrace
378	Goths inflict a heavy defeat on a Roman army at Adrianople in Thrace
388	Theodosius I defeats western usurper Magnus Maximus
394	Theodosius I defeats western usurper Eugenius

406	Vandals and other Germanic groups cross the Rhine on New Year's Eve
429	Vandals cross from Spain to north Africa
439	Vandals capture Carthage
468	Leo's expedition against the Vandals is destroyed
476	Last western emperor (Romulus Augustulus) is deposed by Germanic general Odoacer
502	Persian king Kavad invades Roman territory in northern Mesopotamia and captures Amida
531	Roman forces under Belisarius defeated by Persians at Callinicum
533	Roman general Belisarius invades Vandal north Africa at Justinian's behest and completes the overthrow of the Vandal regime the following year
535	Belisarius invades Sicily to begin campaign to reconquer Italy from the Goths
540	Persian king Khusro I invades Syria, sacking many cities and extorting booty
552	Roman general Narses defeats Gothic king Totila at battle of Taginae in Italy, marking the end of major hosilities in the Gothic war
586	Roman victory over Persian forces at Solachon in northern Mesopotamia
602	Military revolt on the lower Danube results in the overthrow of Maurice and his replacement by Phocas
603	Persian king Khusro II invades the empire
610	Governor Heraclius overthrows Phocas
626	Joint Persian-Avar siege of Constantinople eventually fails
627	Heraclius defeats Persians at Nineveh and imposes peace the following year
636	Arab army defeats Roman army at the Yarmuk in Syria

Roman Emperors

This list presents the names and dates of emperors referred to in this volume. Overlapping dates reflect joint rule, which became increasingly common in the third to fifth centuries.

First Century

Augustus	27 BC–AD 14
Tiberius	14–37
Gaius	37–41
Claudius	41–54
Nero	54–68
Galba	68–69
Otho	69
Vitellius	69
Vespasian	69–79
Titus	79–81
Domitian	81–96
Nerva	96–98
Trajan	98–117

Second Century

Hadrian	117–138
Antoninus Pius	138–161
Marcus Aurelius	161–180
Lucius Verus	161–169
Commodus	180–192
Pertinax	193
Pescennius Niger	193–194
Septimius Severus	193–211
Caracalla	198–217

Third Century

Elagabalus	218–222
Severus Alexander	222–235
Maximinus	235–238
Gordian III	238–244
Philip	244–249
Decius	249–251
Valerian	253–260
Gallienus	253–268
Aurelian	270–275
Probus	276–282
Carus	282–283
Diocletian	284–305
Maximian	286–305
Constantius I (tetrarch)	293–306
Galerius (tetrarch)	293–311

Fourth Century

Maxentius	306–312
Constantine I	306–337
Licinius	308–324
Constans	337–350
Constantius II	337–361
Magnentius (usurper)	350–353
Julian	361–363
Jovian	363–364
Valentinian I	364–375
Valens	364–378
Gratian	375–383
Valentinian II	375–392
Theodosius I	379–395
Arcadius	395–408
Honorius	395–423

Fifth Century

Theodosius II	408–450
Valentinian III	425–455
Marcian	450–457

Leo	457–474
Zeno	474–491
Anastasius	491–518

Sixth Century

Justin I	518–527
Justinian	527–565
Justin II	565–578
Tiberius II	578–582
Maurice	582–602

Seventh Century

Phocas	602–610
Heraclius	610–641

Glossary

assidui Roman citizens who owned sufficient property to qualify for service in the legions (Republic)

auxilia military units recruited from non-citizens in the provinces, with skills (e.g., cavalry, light infantry, archery) that complemented the heavy infantry of the legions (Principate); elite infantry units (Late Antiquity)

auxilia externa military units levied from outside Italy during the later Republic

census registration of Roman citizens and their property, for purposes of classification including liability for military service; conducted by senior magistrates (censors) during the Republic approximately every five years

centuriate assembly one of the citizen assemblies during the Republic, responsible for electing senior magistrates and voting on war and peace; voting rights were determined by census classification on the basis of property

century important sub-unit of the legion (Republic and Principate); commanded by a centurion and usually comprising eighty men

cohort tactical unit of the legion which superseded the maniple in the late Republic; it usually comprised six centuries (so 480 men), with ten cohorts making up a legion; its etymology is uncertain but may relate to *hortus* (garden) and by extension to an area within military camps; also a term used during the Principate for some units of *auxilia*

comitatenses category of military unit which 'accompanied' the emperor, hence troops in mobile field armies, as distinct from *limitanei* (fourth century onwards)

commilito fellow soldier (pl. *commilitones*)

conquisitor recruiting officer (lit. 'seeker out') appointed by a magistrate to levy troops (Republic)

consul senior magistrate in the Republic; two were elected annually and usually commanded an army during their time in office; during the Principate and Late Antiquity, the office became honorific with emperors choosing incumbents (often themselves), but it retained great prestige

contubernalis comrade-in-arms (lit. 'tent mate') (pl. *contubernales*)

contubernium collective term for eight *contubernales* who shared a tent, messed and fought together

denarius silver coin, equivalent to four *sestertii*; a legionary's pay in the later first century BC was 225 *denarii* p.a.

Feriale Duranum papyrus document from the fort at Dura-Europos preserving part of a calendar of religious festivals (*feriae*) observed by the local military unit

genius guardian spirit

limitanei category of military unit stationed in frontier provinces (*limites*) (fourth century onwards)

maniple tactical unit of the legion comprising two centuries, with thirty maniples making up a legion (Republic, 4th to 2nd century BC)

optio a junior officer chosen by a centurion to assist him

pilum heavy javelin (principal missile weapon of legionaries during Republic and Principate)

Principate modern term for the autocratic regime established by Augustus in the late first century BC and continuing broadly to the early third century, in which the emperor was presented as the 'first man' (*princeps*)

proletarii Roman citizens who did not own sufficient property to qualify for service in the legions and whose only contribution to the state was their offspring (*proles*)(also referred to as *capite censi* – those registered by their head, rather than their property)

Republic Roman political system (*res publica*) in place from late sixth century to late first century BC, with power exercised by a combination of citizen assemblies, magistrates and senate (sometimes referred to by the acronym SPQR, i.e. *Senatus Populusque Romanus* ('the senate and people of Rome'))

sacramentum military oath of allegiance

senate originally a body of about three hundred former and current magistrates which, although formally only advisory in capacity, came to exercise great influence in Republican politics; that influence was restricted by emperors, but since senators were generally also very wealthy (there was a property qualification of 1 million HS) and formed the elite of Roman society, they remained important

sestertius coin equivalent to a quarter of a *denarius* (pl. *sestertii*; (abbreviation HS); a legionary's pay in the later first century BC was 900 HS p.a.

Social War the war between Rome and its Italian allies (*socii*) (91–89 BC)

solidi stable gold coin introduced by Constantine in the early fourth century; 3 *solidi* is thought to have been enough to live on for a year; 72 *solidi* = 1 lb of gold

stipendium military pay

Tetrarchy modern term for the arrangement established by Diocletian in 293 whereby rule of the empire was shared between four men; the last vestiges disappeared when Constantine established himself as sole ruler of the empire in 324

tributum tax on Roman citizens introduced in late fifth or fourth century BC to cover the cost of military pay; suspended in 167 BC after the acquisition of Macedonian treasury

References

Adams, J. N. (1994) 'Latin and Punic in contact? The case of the Bu Njem ostraca', *JRS* 84: 87–112

(1999) 'The poets of Bu Njem: Language, culture and the centurionate', *JRS* 89: 109–34

(2003) *Bilingualism and the Latin Language*. Cambridge

Alföldy, G. (1995) 'Eine Bauinschrift aus dem Colosseum', *ZPE* 109: 195–226

Allison, P. (2011) 'Soldiers' families in the early Roman empire' in Rawson, B., ed., *A Companion to Families in the Greek and Roman Worlds*, 160–82. Chichester

(2013) *People and Spaces in Roman Military Bases*. Cambridge

Allmand, C. (2011) *The* De Re Militari *of Vegetius: The Reception, Transmission and Legacy of a Roman Text in the Middle Ages*. Cambridge

Alston, R. (1995) *Soldier and Society in Roman Egypt: A Social History*. London

(1998) 'Arms and the man: Soldiers, masculinity and power in Republican and Imperial Rome' in Foxhall, L., & Salmon, J., eds., *When Men Were Men: Masculinity, Power and Identity in Classical Antiquity*, 205–23. London

(1999) 'The ties that bind: soldiers and societies' in Goldsworthy, A., & Haynes, I., eds., *The Roman Army as a Community*, 175–96. Portsmouth, RI

Ando, C. (2000) *Imperial Ideology and Provincial Loyalty in the Roman Empire*. Berkeley

(2016) 'Triumph in the decentralized empire' in Goldbeck & Wienand 2016: 397–418

Anson, E. (2010) 'The general's pre-battle exhortation in Graeco-Roman warfare', *G&R* 57: 304–18

Arena, V. (2012) Libertas *and the Practice of Politics in the Late Roman Republic*. Cambridge

Ash, R. (1999) *Ordering Anarchy: Armies and Leaders in Tacitus'* Histories. London

Bagnall, R. S. (1993) *Egypt in Late Antiquity*. Princeton

(1997) 'A kinder, gentler Roman army?' *JRA* 10: 504–12

Bagnall, R. S., ed. (2009) *The Oxford Handbook of Papyrology*. Oxford

Bagnall, R. S., & Palme, B. (1996) 'Franks in sixth-century Egypt', *Tyche* 11: 1–10

Bakker, L. (1993) 'Raetien unter Postumus – Das Siegesdenkmal einer Juthungenschlacht im Jahre 260 n. Chr. aus Augsburg', *Germania* 71: 369–86
Balsdon, J. P. V. D. (1951) 'Sulla Felix', *JRS* 41: 1–10
Bang, P. F. (2013) 'Predation' in Scheidel, W., ed., *The Cambridge Companion to the Roman Economy*, 197–217. Cambridge
Barnes, T. D. (1985) 'The career of Abinnaeus', *Phoenix* 39: 368–74
Bartman, E. (2005) 'The mock face of battle', *JRA* 18: 99–119
Batstone, W. W., & Damon, C. (2006) *Caesar's Civil War*. Oxford
Beard, M. (2007) *The Roman Triumph*. Cambridge, MA
Beard, M., North, J., & Price, S. (1998) *Religions of Rome*, 2 vols. Cambridge
Bell, H. I., Martin, V., Turner, E. G., & van Berchem, D., eds. (1962) *The Abinnaeus Archive: Papers of a Roman Officer in the Reign of Constantius II*. Oxford
Bellinger, A. R. (1966) *Catalogue of the Byzantine Coins in the Dumbarton Oaks Collection and in the Whittemore Collection. Vol. 1: Anastasius to Maurice, 491–602*. Washington, DC
Bennett, J. (2000) *Trajan: Optimus Princeps*. London
Beyerchen, A. (1992) 'Clausewitz, nonlinearity and the unpredictability of war', *International Security* 17: 59–90
Birley, A. R. (1983) 'The economic effects of Roman frontier policy' in King, A., & Henig, M., eds., *The Roman West in the Third Century*, 39–53. Oxford
(1999) *Septimius Severus*, 2nd ed. London
(2000a) *Marcus Aurelius: A Biography*, 2nd ed. London
(2000b) 'The life and death of Cornelius Tacitus', *Historia* 49: 230–47
Birley, R. (2009) *Vindolanda: A Roman Frontier Fort on Hadrian's Wall*. Stroud
Bishop, M. C., & Coulston, J. C. N. (2006) *Roman Military Equipment from the Punic Wars to the Fall of Rome*, 2nd ed. Oxford
Blockley, R. C. (1975) *Ammianus Marcellinus: A Study of his Historiography and Political Thought*. Brussels
Börm, H. (2013) 'Justinians Triumph und Belisars Erniedrigung. Überlegungen zum Verhältnis zwischen Kaiser und Militär im späten Römischen Reich', *Chiron* 43: 63–91
Bosworth, A. B. (2003) '*Plus ça change* ... Ancient historians and their sources', *Classical Antiquity* 22: 167–98
Bowersock, G. W. (1978) *Julian the Apostate*. London
Bowman, A. K. (1994a) *Life and Letters on the Roman Frontier: Vindolanda and Its People*. London
(1994b) 'The Roman imperial army: letters and literacy on Rome's northern frontier' in Bowman, A. K., & Woolf, G., eds., *Literacy and Power in the Ancient World*, 109–25. Cambridge
Bradley, K. (1998) 'The sentimental education of the Roman child: the role of pet-keeping', *Latomus* 57: 523–57
(2004) 'On captives under the Principate', *Phoenix* 58: 298–318
Brennan, P. (1998a) 'Divide and fall: the separation of legionary cavalry and the fragmentation of the Roman empire', in Hillard, T. W., Kearsley, R. A.,

Nixon, C. E. V., Nobbs, A. M., eds., *Ancient History in a Modern University. Vol. 2: Early Christianity, Late Antiquity, and Beyond*, 238–44. Grand Rapids, MI, and Cambridge
(1998b) 'The last of the Romans: Roman identity and the Roman army in the late Roman Near East', *Mediterranean Archaeology* 11: 191–203
(1998c) 'The user's guide to the *Notitia Dignitatum*: the case of the *dux Armeniae* (*ND Or.* XXXVIII)', *Antichthon* 32: 34–49
(2007) 'Zosimus 2.34.1 and "the Constantinian reform"' in Lewin & Pellegrini 2007: 211–18
Brice, L. L. (2003) 'Holding a wolf by the ears: Mutiny and unrest in the Roman military, 44 BC–AD 68.' Diss. University of North Carolina at Chapel Hill
(2011) 'Disciplining Octavian: A case study of Roman military culture, 44–30 BCE' in Lee, W. E., ed., *Warfare and Culture in World History*, 35–60. New York
(2015) 'Second chance for valor: Restoration of order after mutinies and indiscipline in the Roman military' in Brice, L. L., & Slootjes, D., eds., *Ancient World Views: Institutions and Geography from the Greco-Roman World*, 103–21. Leiden
Brice, L. L., & Roberts, J. T., eds. (2011) *Recent Directions in the Military History of the Ancient World*. Claremont, CA
Brody, L. R., & Hoffman, G. L., eds. (2014) *Roman in the Provinces: Art on the Periphery of Empire*. Chestnut Hill, MA
Broughton, T. R. S. (1938) *Roman Asia: An Economic Survey of Ancient Rome, Vol. IV*, ed. T. Frank. Baltimore
Brouquier-Reddé, V. (1992) *Temples et cultes de Tripolitanie*. Paris
Brown, T. S. (1984) *Gentlemen and Officers: Imperial Administration and Aristocratic Power in Byzantine Italy, AD 554–800*. Rome
Brunt, P. A. (1962) 'The army and the land in the Roman revolution', *JRS* 52: 69–86 (reprinted in Brunt 1988: 240–80 with revisions)
(1971) *Social Conflicts in the Roman Republic*. London
(1974) 'Conscription and volunteering in the Roman imperial army', *Scripta Classica Israelica* 1: 90–115 (reprinted in Brunt 1990: 188–214, with Addenda (512–13))
(1987) *Italian Manpower, 225 BC–AD 14*. Oxford (reprint of 1971 edition with additional Postscript at 717–27)
(1988) *The Fall of the Roman Republic and Related Essays*. Oxford
(1990) *Roman Imperial Themes*. Oxford
Bruun, C., & Edmondson, J., eds. (2014) *The Oxford Handbook of Roman Epigraphy*. Oxford
Burns, T. S. (1973) 'The battle of Adrianople: a reconsideration', *Historia* 22: 336–45
Callu, J. P. (1995) 'Le butin de Neupotz', *JRA* 8: 514–20
Cameron, A. (2011) *The Last Pagans of Rome*. Oxford
Campbell, B. (1984) *The Emperor and the Roman Army, 31 BC–AD 235*. Oxford
(1987) 'Teach yourself how to be a general', *JRS* 77: 13–29
(1994) *The Roman Army, 31 BC–AD 337: A Sourcebook*. London

(2002) *War and Society in Imperial Rome, 31 BC–AD 284*. London
(2004) *Greek and Roman Military Writers: Selected Readings*. London
Campbell, B., & Tritle, L. A., eds. (2013) *The Oxford Handbook of Warfare in the Classical World*. Oxford
Carrié, J.-M. (1986) 'L'esercito: trasformazioni funzionali ed economie locali' in Giardina. A., ed., *Società romana e impero tardoantica: istituzioni, ceti, economie*, 449–88. Rome
(1993) 'The soldier' in Giardina, A., ed., *The Romans*, tr. L. G. Cochrane, 100–37. Chicago
Carroll, M. (2005) 'The preparation and consumption of food as a contributing factor towards communal identity in the Roman army' in Visy, Z., ed., *Limes XIX: Proceedings of the XIXth International Congress of Roman Frontier Studies held in Pécs, Hungary, September 2003*, 363–72. Pécs
Champlin, E. (1991) *Final Judgements: Duty and Emotion in Roman Wills, 20 BC–AD 250*. Berkeley
Chaniotis, A. (2005) *War in the Hellenistic World: A Social and Cultural History*. Malden and Oxford
Charles, M. B. (2007) *Vegetius in Context: Establishing the Date of the* Epitoma Rei Militaris. Stuttgart
(2010) 'Unseemly professions and recruitment in late antiquity: *Piscatores* and Vegetius *Epitoma* 1.7.1–2', *AJP* 131: 101–20
Cherf, W. J. (1993) 'The Thermopylae garrison of *Vita Claudii* 16', *CPh* 88: 230–36
Cherry, D. (1998) *Frontier and Society in Roman North Africa*. Oxford
Chlup, J. T. (2014) 'Just war in Onasander's Strategikos', *Journal of Ancient History* 2: 37–63
Chrissanthos, S. G. (1997) 'Scipio and the mutiny at Sucro, 206 BC', *Historia* 46: 172–84
(2001) 'Caesar and the mutiny of 47 BC', *JRS* 91: 63–75
(2004) 'Freedom of speech and the Roman Republican army' in Sluiter, I., & Rosen, R., eds., *Free Speech in Classical Antiquity*, 341–67. Leiden
(2013) 'Keeping military discipline' in Campbell & Tritle 2013: 312–29
Clackson, J. (2015) *Language and Society in the Greek and Roman Worlds*. Cambridge
Clark, A. J. (2007) *Divine Qualities: Cult and Community in Republican Rome*. Oxford
Clark, J. C. (2014) *Triumph in Defeat: Military Loss and the Roman Republic*. Oxford
Clauss, M. (1986) 'Heerwesen (Heeresreligion)', *RAC* 17: 1073–113
Collar, A. (2013) *Religious Networks in the Roman Empire: The Spread of New Ideas*. Cambridge
Conant, J. (2012) *Staying Roman: Conquest and Identity in Africa and the Mediterranean, 439–700*. Cambridge
Cool, H. E. M. (2004) *The Roman Cemetery at Brougham, Cumbria: Excavations 1966–67*. London

Cooley, A. E. (2012a) *The Cambridge Manual of Latin Epigraphy.* Cambridge
(2012b) 'Commemorating the war dead of the Roman world' in Low, P., Oliver, G., & Rhodes, P.J., eds., *Cultures of Commemoration: War Memorials, Ancient and Modern*, 63–88. Oxford
Corbier, M. (1974) *L'*aerarium Saturni *et l'*aerarium militare: *Administration et prosopographie sénatoriale.* Rome
Cornell, T. (1993) 'The end of Roman imperial expansion' in Rich & Shipley 1993: 139–70
(1995) *The Beginnings of Rome: Italy and Rome from the Bronze Age to the Punic Wars (c.1000–264 BC).* London
(1996) 'Hannibal's legacy: the effects of the Hannibalic War on Italy' in Cornell et al. 1996: 97–117
Cornell, T., Rankov, B., & Sabin, P., eds. (1996) *The Second Punic War: A Reappraisal.* London
Cornwell, H. (2017) Pax *and the Politics of Peace: Republic to Principate.* Oxford
Cosme, P. (1993) 'Le livret militaire du soldat romain', *Cahiers du Centre Gustave Glotz* 4: 67–80
Coulston, J. (1985) 'Roman archery equipment' in Bishop, M. C., ed., *The Production and Distribution of Roman Military Equipment*, 220–366. Oxford
(1990) 'Late Roman armour, third-sixth centuries', *Journal of Roman Military Equipment Studies* 1: 139–60
(2003) 'Tacitus *Historiae* I.79 and the impact of Sarmatian warfare on the Roman empire' in von Carnap-Bornheim, C., ed., *Kontakt – Kooperation – Konflikt: Germanen und Sarmaten zwischen dem 1. und 4. Jahrhundert nach Christus*, 415–33. Marburg
(2007) 'By the sword united: Roman fighting styles on the battlefield and in the arena' in Molloy, B., ed., *The Cutting Edge: Studies in Ancient and Medieval Combat*, 34–51. Stroud
(2013) 'Courage and cowardice in the Roman imperial army', *War in History* 20: 7–31
(2018) 'The army in imperial Rome' in Holleran, C., & Claridge, A., eds., *A Companion to the City of Rome*, 173–95. Chichester
Crawford, M. (1978) *The Roman Republic.* London
Crowley, J. (2015) 'Beyond the universal soldier: Combat trauma in classical antiquity' in Meineck, P., & Konstan, D., eds., *Combat Trauma and the Ancient Greeks*, 111–30. London
Culham, P. (1989) 'Chance, command, and chaos in ancient military engagements', *World Futures* 27: 191–205
Cuvigny, H., ed. (2003) *La route de Myos Hormos: l'armée romaine dans le désert oriental d'Egypte.* Cairo
Daly, G. (2002) *Cannae: The Experience of Battle in the Second Punic War.* London
Damon, C. (2007) 'Rhetoric and historiography' in Dominik, W., & Hall, J., eds., *A Companion to Roman Rhetoric*, 439–50. Chichester

Darby, R. (2015) 'The late Roman military baths of the Wadi Arabah: a survey of recent archaeological work', *Syria* 92: 67–83

Dart, C. J. (2014) *The Social War, 91 to 88 BCE: A History of the Italian Insurgency against the Roman Republic*. London

Davenport, C. (2018) *A History of the Roman Equestrian Order*. Cambridge

de Blois, L., & Lo Cascio, E., eds. (2007) *The Impact of the Roman Army (200 BC–AD 476): Economic, Social, Political, Religious and Cultural Aspects* (= *Impact of Empire* vol. 6). Leiden

de Ligt, L. (2012) *Peasants, Citizens and Soldiers: Studies in the Demographic History of Roman Italy, 225 BC–AD 100*. Cambridge

(2015) 'Civilians: Republic' in Le Bohec 2015: 235–6

de Ligt, L., & Northwood, S., eds. (2008) *People, Land and Politics: Demographic Developments and the Transformation of Roman Italy, 300 BC–AD 14*. Leiden

Dench, E. (2005) *Romulus' Asylum: Roman Identities from the Age of Alexander to the Age of Hadrian*. Oxford

(2018) *Empire and Political Cultures in the Roman World*. Cambridge

Dennis, G. T. (1981) 'Byzantine battle flags', *Byzantinische Forschungen* 8: 51–9

(1984) *Maurice's Strategikon: Handbook of Byzantine Military Strategy*. Philadelphia

Dillon, S. (2006) 'Women on the Columns of Trajan and Marcus Aurelius and the visual language of Roman victory' in Dillon & Welch 2006: 244–71

Dillon, S., & Welch, K. E., eds. (2006) *Representations of War in Ancient Rome*. Cambridge

Dirven, L. (1999) *The Palmyrenes of Dura-Europos*. Leiden

Dobson, B. (1986) 'The Roman army: wartime or peacetime army?' in Eck, W., & Wolff, H., eds., *Heer und Integrationspolitik: Die römischen Militärdiplome als historische Quelle*, 10–25. Cologne

Dodgeon, M. H., & Lieu, S. N. C. (1991) *The Roman Eastern Frontier and the Persian Wars, AD 226–363: A Documentary History*. London

Drew-Bear, T. (1981) 'Les voyages d'Aurélius Gaius, soldat de Dioclétien', in *La géographie administrative et politique d'Alexandre à Mahomet: Actes du Colloque de Strasbourg 14–16 juin 1979*, 93–141. Strasbourg

Drinkwater, J. F. (2000) 'The revolt and ethnic origin of the usurper Magnentius (350–353) and the rebellion of Vetranio' *Chiron* 30: 131–59

Drinkwater, J. F., & Lee, A. D. (2014) 'Civil wars: Late Empire' in Le Bohec 2014: 211–18

Duncan-Jones, R. P. (1994) *Money and Government in the Roman Empire*. Cambridge

(1996) 'The impact of the Antonine plague', *JRA* 9: 108–93

(2016) *Power and Privilege in Roman Society*. Cambridge

Dyson, S. (1971) 'Native revolts in the Roman empire', *Historia* 20: 239–74

Eadie, J. W. (1967) 'The development of Roman mailed cavalry', *JRS* 57: 161–73

Eck, W. (2011) 'Septimius Severus und die Soldaten: das Problem der Soldatenehe und ein neues Auxiliardiplom' in Onken, B., & Rohde, D., eds., *In*

omni historia curiosus. Studien zur Geschichte von der Antike bis zur Neuzeit: Festschrift für Helmuth Schneider zum 65. Geburtstag, 63–77. Wiesbaden
(2014) '*Milites et pagani*: La posizione dei soldati nella società romana', *Rationes Rerum: Rivista di filologia e storia* 3: 11–54
Eck, W., Caballos, A., & Fernández, F. (1996) *Das senatus consultum de Cn. Pisone patre*. Munich
Eckardt, H. (2018) *Writing and Power in the Roman World: Literacies and Material Culture*. Cambridge
Eckstein, A. M. (2006) *Mediterranean Anarchy, Interstate War, and the Rise of Rome*. Berkeley
Eich, A., & Eich, P. (2005) 'War and state-building in Roman Republican times', *SCI* 24: 1–33
Elton, H. (1996) *Warfare in Roman Europe, AD 350–425*. Oxford
(2007) 'The Later Roman Empire: Military forces' in Sabin, Van Wees & Whitby 2007: vol. 2, 270–309
(2018) *The Roman Empire in Late Antiquity: A Political and Military History*. Cambridge
Engen, R. (2009) *Canadians Under Fire: Infantry Effectiveness in the Second World War*. Montreal
(2011) 'S. L. A. Marshall and the ratio of fire: History, interpretation, and the Canadian experience', *Canadian Military History* 20: 39–48
Erdkamp, P. (1998) *Hunger and the Sword: Warfare and Food Supply in Roman Republican Wars (264–30 BC)*. Amsterdam
Erdkamp, P., ed. (2002) *The Roman Army and the Economy*. Amsterdam
ed. (2007) *A Companion to the Roman Army*. Chichester
Fears, J. R. (1981) 'The theology of victory at Rome: Approaches and problems', *ANRW* 17.2: 736–826
Fentress, E. (1979) *Numidia and the Roman Army: Social, Military and Economic Aspects of the Frontier Zone*. Oxford
(1983) 'Forever Berber?', *Opus* 2: 161–75
(2005) 'Toynbee's legacy: Southern Italy after Hannibal', *JRA* 18: 482–8
Fink, R. O. (1971) *Roman Military Records on Papyrus*. Cleveland
Fink, R. O., Hoey, A. S., & Snyder, R. F. (1940) 'The *Feriale Duranum*', *YCS* 7: 1–222
Fisher, G. (2011) *Between Empires: Arabs, Romans and Sasanians in Late Antiquity*. Oxford
Fishwick, D. (1988) 'Dated inscriptions and the *Feriale Duranum*', *Syria* 65: 349–61
Flower, H. I. (2000) 'The tradition of the *spolia opima*: M. Claudius Marcellus and Augustus', *CA* 19: 34–64
Fowden, G. (1988) 'City and mountain in late Roman Attica', *JHS* 108: 48–59
Freeman, P., & Kennedy, D., eds. (1986) *The Defence of the Roman and Byzantine East*. Oxford
Frend, W. H. C. (1956) 'A third-century inscription relating to *angareia* in Phrygia', *JRS* 46: 46–56

Frier, B. (2000) 'Demography' in *CAH² XI: The High Empire, AD 70–192*: 787–816
Fronda, M. P. (2010) *Between Rome and Carthage: Southern Italy in the Second Punic War*. Cambridge
Fuhrmann, C. J. (2012) *Policing the Roman Empire: Soldiers, Administration and Public Order*, Oxford
Fulkerson, L. (2013) *No Regrets: Remorse in Classical Antiquity*. Oxford
Funari, P. P. A. (2002) 'The consumption of olive oil in Roman Britain and the role of the army' in Erdkamp 2002: 235–63
Gagé, J. (1933a) 'La théologie de la victoire impériale', *Revue historique* 171: 1–43
(1933b) '*Stauros nikopoios*: La victoire impériale dans l'empire chrétien', *Revue d'histoire et de philosophie religieuses* 13: 370–400
Galinsky, K. (1996) *Augustan Culture: An Interpretive Introduction*. Princeton
Garnsey, P. (1970) 'Septimius Severus and the marriage of soldiers', *California Studies in Classical Antiquity* 3: 45–53
(1988) *Famine and Food Supply in the Graeco-Roman World: Responses to Risk and Crisis*. Cambridge
(1994) *Food and Society in Classical Antiquity*. Cambridge
(2004) 'Roman citizenship and Roman law in the Late Empire' in Swain & Edwards 2004: 133–55
(2019) 'The late Roman army', in Maiuro, M., Merola, G. D., de Nardis, M., Soricelli, G., eds., *Studi di Storia per Elio Lo Cascio*, 321–41. Bari
Gilliver, K. (1996) 'The Roman army and the morality of war' in Lloyd 1996: 219–38
(2007) 'The Late Republic and the Principate: Battle' in Sabin et al. 2007: vol. 2, 122–57
Gleason, M. W. (2009) 'Defending Republican male *virtus*', *JRA* 22: 474–6
Glenn, R. W. (2000) *Reading Athena's Dance Card: Men Against Fire in Vietnam*. Annapolis
Goffman, E. (1961) *Asylums: Essays on the Social Situation of Mental Patients and Other Inmates*. Chicago
Goldbeck, F., & Wienand, J., eds. (2016) *Der Römische Triumph in Prinzipat und Spätantike*. Berlin
Goldstein, J. S. (2001) *War and Gender: How Gender Shapes the War System and Vice Versa*. Cambridge
Goldsworthy, A. (1996) *The Roman Army at War, 100 BC–AD 200*. Oxford
Goodman, M. (1987) *The Ruling Class of Judaea: The Origins of the Jewish Revolt against Rome, AD 66–70*. Cambridge
(2008) *Rome and Jerusalem: The Clash of Ancient Civilisations*. London
Gordon, R. L. (2009) 'The Roman army and the cult of Mithras: a critical review' in Wolff 2009: 379–450
Gowing, A. M. (1992) *The Triumviral Narratives of Appian and Cassius Dio*. Ann Arbor
Gray, C. (2015) *Jerome: Vita Malchi*. Oxford
Gray, G. (1959) *The Warriors: Reflections on Men in Battle*. New York

Greatrex, G. (1998) *Rome and Persia at War, 502–532*. Leeds
(2000) 'Roman identity in the sixth century' in Mitchell, S., & Greatrex, G., eds., *Ethnicity and Culture in Late Antiquity*, 267–92. London
(2001) 'Justin I and the Arians', *Studia Patristica* 34: 72–81
Greene, E. M. (2015) '*Conubium cum uxoribus*: wives and children in the Roman military diplomas', *JRA* 28: 125–59
Gregory, S. (1995) *Roman Military Architecture on the Eastern Frontier from AD 200–600*, 3 vols. Amsterdam
Griffin, M., ed. (2009) *A Companion to Julius Caesar*. Chichester
Grosse, R. (1920) *Römische Militärgeschichte von Gallienus bis zum Beginn der byzantinischen Themenverfassung*. Berlin
Haarer, F. (2006) *Anastasius I: Politics and Empire in the Late Roman World*. Cambridge
Haeckel, A. E. (1987) 'The *principia* of el-Lejjūn' in Parker, S. T., ed., *The Roman Frontier in Central Jordan: Interim Report on the Limes Arabicus Project, 1980–1985*, 203–60. Oxford
Haensch, R. (2004) 'La christianisation de l'armée romaine' in Le Bohec, Y., & Wolff, C., eds., *L'Armée romaine de Dioclétien à Valentinien Ier*, 525–31. Lyons
(2012) 'The Roman army in Egypt' in Riggs, C., ed., *The Oxford Handbook of Roman Egypt*, 68–82. Oxford
Haldon, J. (2016) *The Empire That Would Not Die: The Paradox of Eastern Roman Survival, 640–740*. Cambridge, MA
et al. (2018) 'Plagues, climate change, and the end of an empire. A response to Kyle Harper's *The Fate of Rome*' [3 parts], *History Compass* 16, e12508, e12506, e12507
Halkin, L. (1953) *La supplication d'action de graces chez les Romains*. Paris
Hammond, N. G. L. (1988) 'The campaign and the battle of Cynoscephalae, 197 BC', *JHS* 108: 60–82
Hansen, M. H. (1993) 'The battle exhortation in ancient historiography: fact or fiction?', *Historia* 42: 161–80
(1998) 'The little grey horse – Henry V's speech at Agincourt and the battle exhortation in ancient historiography', *Histos* 2: 46–63
Hanson, V. D. (2000) *The Western Way of War: Infantry Battle in Classical Greece*, new ed. Berkeley
Hanson, W. S. & Keppie, L. J. F., eds. (1980) *Roman Frontier Studies 1979*. Oxford
Hardwick, L. (2000) 'Concepts of peace' in Huskinson, J., ed., *Experiencing Rome: Power, Culture and Identity in the Roman Empire*, 335–68. London
Harmand, J. (1967) *L'Armée et le soldat à Rome de 107 à 50 avant notre ère*. Paris
Harper, K. (2011) *Slavery in the Late Roman World, AD 275–425*. Cambridge
(2015) 'Pandemics and passages to late antiquity: Rethinking the plague of *c.*249–270 described by Cyprian', *JRA* 28: 233–60
(2017) *The Fate of Rome: Climate, Disease and the End of an Empire*. Princeton

Harris, W. V. (1979) *War and Imperialism in Republican Rome, 327–70 BC.* Oxford (reprinted with new introduction 1984)
Harris, W. V., ed., (1984a) *The Imperialism of Mid-Republican Rome.* Rome
(1984b) 'The Italians and the empire' in Harris 1984a: 89–109
(1989) *Ancient Literacy.* Cambridge, MA
(2005) 'Constantine's dream', *Klio* 87: 488–94
(2006) 'Readings in the narrative literature of Roman courage' in Dillon & Welch 2006: 300–20
(2007) 'The late Republic' in Scheidel et al. 2007: 511–39
(2009) *Dreams and Experience in Classical Antiquity.* Cambridge, MA
(2016) *Roman Power: A Thousand Years of Empire.* Cambridge
Hassall, M. W. (2000) 'The army' in *CAH*[2] *XI: The High Empire, AD 70–192*: 320–343
Hauken, T. (1998) *Petition and Response: An Epigraphic Study of Petitions to Roman Emperors 181–249.* Bergen
Hauken, T., & Malay, H. (2009) 'A new edict of Hadrian from the province of Asia setting regulations for requisitioned transport' in Haensch, R., ed., *Selbstdarstellung und Kommunikation: Die Veröffentlichung staatlicher Urkunden auf Stein und Bronze in der römischen Welt*, 327–48. Munich
Haynes, I. (2013) *Blood of the Provinces: The Roman* Auxilia *and the Making of Provincial Society from Augustus to the Severans.* Oxford
Heather, P. J. (1991) *Goths and Romans*, 332–489. Oxford
(2005) *The Fall of the Roman Empire: A New History.* London
(2018) *Rome Resurgent: War and Empire in the Age of Justinian.* New York
Hebblewhite, M. (2017) *The Emperor and the Army in the Later Roman Empire, AD 235–395.* London
Hendy, M. (1985) *Studies in the Byzantine Monetary Economy, c.300–1450.* Cambridge
Hin, S. (2013) *The Demography of Roman Italy: Population Dynamics in an Ancient Conquest Society, 201 BCE–14 CE.* Cambridge
Hirt, A. M. (2010) *Imperial Mines and Quarries in the Roman World: Organizational Aspects, 27 BC–AD 235.* Oxford
Hoff. M. C. (1997) '*Laceratae Athenae*: Sulla's siege of Athens in 87/6 BC and its aftermath' in Hoff, M. C., & Rotroff, S. I., eds., *The Romanization of Athens*, 33–51. Oxford
Hoffmann, D. (1969–70) *Das spätrömische Bewegungsheer und die* Notitia Dignitatum, 2 vols. Düsseldorf
Holmes, R. (1985) *Acts of War: The Behaviour of Men in Battle.* New York
Hölscher, T. (1967) *Victoria Romana.* Mainz
Hope, V. M. (2001) *Constructing Identity: The Roman Funerary Monuments of Aquileia, Mainz and Nîmes.* Oxford
(2003) 'Trophies and tombstones: commemorating the Roman soldier', *World Archaeology* 35: 79–97
Hopkins, K. (1978) *Conquerors and Slaves. Sociological Studies in Roman History I.* Cambridge

(1999) *A World Full of Gods: Pagans, Jews and Christians in the Roman Empire.* London
(2002) 'Rome, taxes, trade and rents' in Scheidel & von Reden 2002: 190–230
Horsfall, N. (2003) *The Culture of the Roman Plebs.* London
Howard-Johnston, J. (2010) *Witnesses to a World Crisis: Historians and Histories of the Middle East in the Seventh Century.* Oxford
Hoyos, D., ed. (2011) *A Companion to the Punic Wars.* Chichester
Humphries, M. (2008) 'From usurper to emperor: the politics of legitimation in the age of Constantine', *JLA* 1: 82–100
Isaac, B. (1988) 'The meaning of the terms *limes* and *limitanei*', *JRS* 78: 125–47
(1992) *The Limits of Empire: The Roman Army in the East*, rev'd ed. Oxford
(1998) *The Near East under Roman Rule: Selected Papers.* Leiden
James, S. (1999) 'The community of the soldiers: A major identity and centre of power in the Roman empire' in Baker, P. *et al.*, eds., *TRAC 1998: Proceedings of the Eighth Annual Theoretical Roman Archaeology Conference, Leicester 1998*, 14–25. Oxford
(2001) '"Romanization" and the peoples of Britain' in Keay & Terrenato 2001: 187–209
(2002) 'Writing the legions: The development and future of Roman military studies in Britain', *Archaeological Journal* 159: 1–58
(2004) *Excavations at Dura-Europos, 1928–1937, Final Report VII (The Arms and Armour and other Military Equipment).* London
(2005) 'The deposition of military equipment during the final siege at Dura-Europos, with particular regard to the Tower 19 countermine', *Carnuntum Jahrbuch 2005*: 198–206
(2006) 'The impact of steppe people and the Partho-Sasanian world on the development of Roman military equipment and dress, 1st to 3rd centuries AD' in Mode, M., & Tubach, J., eds., *Arms and Armour as Indicators of Cultural Transfer: The Steppes and the Ancient World from Hellenistic Times to the Early Middle Ages*, 357–92. Wiesbaden
(2010) 'The point of the sword: what Roman-era weapons could do to bodies – and why they often didn't' in Busch, A. W., & Schalles, H. J., eds., *Waffen in Aktion*, 41–54. Xanten
(2011a) *Rome and the Sword: How Warriors and Weapons Shaped Roman History.* London
(2011b) 'Stratagems, combat, and "chemical warfare" in the siege mines of Dura-Europos', *AJA* 115: 69–101
(2013) 'The archaeology of war' in Campbell & Tritle 2013: 91–127
(2014) 'The "Romanness of the soldiers": barbarized periphery or imperial core?' in Brody & Hoffman 2014: 91–107
Jones, A. H .M. (1964) *The Later Roman Empire 284–602: A Social, Economic and Administrative Survey.* Oxford
Jones, C. P. (1986) *Culture and Society in Lucian.* Cambridge, MA
Jordan, K. C. (2002) 'Right for the wrong reasons: S. L. A. Marshall and the ratio of fire in Korea', *Journal of Military History* 66: 135–62

Kaegi, W. E. (1981) *Byzantine Military Unrest, 471–843*. Amsterdam
(2003) *Heraclius, Emperor of Byzantium*. Cambridge
Kagan, K. (2006) *The Eye of Command*. Ann Arbor
Kay, P. (2014) *Rome's Economic Revolution*. Oxford
Keay, S. (2001) 'Romanization and the Hispaniae' in Keay & Terrenato 2001: 117–44
Keay, S., & Terrenato, N., eds. (2001) *Italy and the West: Comparative Issues in Romanization*. Oxford
Keegan, J. (1976) *The Face of Battle: A Study of Agincourt, Waterloo and the Somme*. Harmondsworth
(1987) *The Mask of Command: A Study of Generalship*. London
Keenan, J. G. (1994) 'Soldier and civilian in Byzantine Hermopolis', *Proceedings of the 20th International Congress of Papyrologists, Copenhagen 23–29 August 1992*, 444–51. Copenhagen
Kelly, G. (2008) *Ammianus Marcellinus, the Allusive Historian*. Cambridge
Kennedy, D. L., ed. (1996) *The Roman Army in the East*. Ann Arbor
Keppie, L. J. F. (1983) *Colonisation and Veteran Settlement in Italy, 47–14 BC*. London
(1984) *The Making of the Roman Army: from Republic to Empire*. London
(1997) 'The changing face of the Roman legions (49 BC–AD 69)', *PBSR* 65: 89–102
Kettenhofen, E. (1994) 'Deportations: ii. In the Parthian and Sasanian periods', *Encyclopedia Iranica*, vol. 7, 297–308. Costa Mesa
King, A. (2001) 'The Romanization of diet in the western empire: comparative zoological studies' in Keay & Terrenato 2001: 210–23
Kneebone, E. (2020) *Oppian's Halieutica: Charting a Didactic Epic*. Cambridge
Kontokosta, A. H. (2013) 'Reconsidering the arches (*fornices*) of the Roman Republic', *JRA* 26: 7–35
Kraeling, C. H. (1967) *The Excavations at Dura-Europos, Final Report VIII, Part II (The Christian Building)*. New Haven
Kraemer, C. J. (1958) *Excavations at Nessana*, vol. 3: *The Non-literary Papyri*. Princeton
Kragelund, P. (2001) 'Dreams, religion and politics in Republican Rome', *Historia* 50: 53–95
Kraus, C. S. (2007) 'Caesar's account of the battle of Massilia (BC 1.34–2.22): Some historiographical and narratological approaches' in Marincola 2007: 371–8
(2009) 'Bellum Gallicum' in Griffin 2009: 159–74
Kulikowski, M. (2007) *Rome's Gothic Wars*. Cambridge
Künzl, E., ed. (1993) *Die Alammanenbeute aus dem Rhein bei Neupotz: Plünderungsgut aus dem römischen Gallien*, 4 vols. Mainz
Lange, C. H. (2009) Res Publica Restituta: *Actium, Apollo and the Accomplishment of the Triumviral Assignment* (= *Impact of Empire* vol. 10). Leiden
(2016) 'The late Republican triumph: continuity and change' in Goldbeck & Wienand 2016: 29–58

Lavan, M. (2016) 'The spread of Roman citizenship, 14–212 CE: Quantification in the face of high uncertainty', *P&P* 230: 3–46
(2019a) 'The foundation of empire? The spread of Roman citizenship from the fourth century BCE to the third century CE' in Berthelot, K., & Price, J., eds., *In the Crucible of Empire: The Impact of Roman Citizenship upon Greeks, Jews and Christians*, 21–54. Leuven
(2019b) 'The army and the spread of Roman citizenship', *JRS* 109
Lazenby, J. F. (1978) *Hannibal's War: A Military History of the Second Punic War*. Warminster
(1996) *The First Punic War: A Military History*. London
Leary, T. J. (1996) *Martial Book XIV: The Apophoreta*. London
Le Bohec, Y. (1994) *The Roman Imperial Army*, tr. R. Bate. London
(2006) *L'Armée romaine sous le Bas-Empire*. Paris
(2010) 'L'écrit au sein de l'armée romaine, du Ier au IIIe siècle de notre ère' in Perrin, Y., ed., *Neroniana VIII*, 192–207. Brussels
Le Bohec, Y., ed. (2015) *The Wiley-Blackwell Encyclopedia of the Roman Army*. Chichester
Lee, A. D. (1996) 'Morale and the Roman experience of battle' in Lloyd 1996: 199–218
(2007) *War in Late Antiquity: A Social History*. Chichester
(2013a) *From Rome to Byzantium, AD 363 to 565: The Transformation of Ancient Rome*. Edinburgh
(2013b) 'Theodosius and his generals' in Kelly, C., ed., *Theodosius II: Rethinking the Roman Empire in Late Antiquity*, 90–108. Cambridge
(2020) 'A professional Roman army?' in Stewart, E., Harris, E., & Lewis, D., eds., *Skilled Labour and Professionalism in Ancient Greece and Rome*, 362–82. Cambridge
Leigh, M. (1995) 'Wounding and popular rhetoric at Rome', *BICS* 40: 195–210
Lendon, J. E. (1997) *Empire of Honour: The Art of Government in the Roman World*. Oxford
(1999) 'The rhetoric of combat: Greek military theory and Roman culture in Julius Caesar's battle descriptions', *Classical Antiquity* 18: 273–329
(2004) 'The Roman army now', *Classical Journal* 99: 141–9
(2005) *Soldiers and Ghosts: A History of Battle in Classical Antiquity*. New Haven
(2006) '*Contubernalis, commanipularis*, and *commilito* in Roman soldiers' epigraphy: drawing the distinction,' *ZPE* 157: 270–76
(2017) 'Battle description in the ancient historians', *G&R* 64: 39–64, 145–67
Lenski, N. (1997) '*Initium mali Romano imperio*: Contemporary reactions to the Battle of Adrianople', *TAPA* 127: 129–68
(2002a) *Failure of Empire: Valens and the Roman State in the Fourth Century AD*. Berkeley
(2002b) 'Were Valentinian, Valens and Jovian confessors before Julian the Apostate?', *ZAC* 6: 253–76
(2007) 'Two sieges of Amida (AD 359 and 502–503) and the experience of combat in the late Roman Near East' in Lewin & Pellgrini 2007: 219–36

(2011a) 'Captivity and Romano-barbarian interchange' in Mathisen, R., & Shanzer, D., eds., *Romans, Barbarians and the Transformation of the Roman World*, 185–98. Farnham

(2011b) 'Captivity and slavery among the Saracens in late antiquity (*ca.* 250–630 CE)', *AnTard* 19: 237–66

Levene, D. (2009) 'Warfare in the *Annals*' in Woodman, A. J., ed., *The Cambridge Companion to Tacitus*, 225–38. Cambridge

(2010) *Livy on the Hannibalic War*. New York

Levick, B. (1990) *Claudius*. London

(2017) *Vespasian*, 2nd ed. London

Levithan, J. (2013) *Roman Siege Warfare*. Ann Arbor

Lewin, A. S., & Pellegrini, P., eds. (2007) *The Late Roman Army in the Near East from Diocletian to the Arab Conquest*. Oxford

Liebeschuetz, J. H. W. G. (1988) 'Ammianus, Julian and divination' in Wissemann, M., ed., *Roma Renascens: Beiträge zur Spätantike und Rezeptionsgeschichte Ilona Opelt gewidmet*, 198–213. Frankfurt

(1990) *Barbarians and Bishops: Army, Church and State in the Age of Arcadius and Chrysostom*. Oxford

Lieu, S. N. C. (1986) 'Captives, refugees and exiles: a study of cross-frontier civilian movements and contacts between Rome and Persia from Valerian to Jovian' in Freeman & Kennedy 1986: 475–505

Linderski, J. (1984) '*Si vis pacem, para bellum*: Concepts of defensive imperialism' in Harris 1984a: 133–52

(2001) 'Silver and gold of valor: the award of *armillae* and *torques*', *Latomus* 60: 3–15

Lintott, A. W. (1999) *The Constitution of the Roman Republic*. Oxford

Lloyd, A. B., ed. (1996) *Battle in Antiquity*. London

Lo Cascio, E. (2007) 'The early Roman empire: the state and the economy' in Scheidel et al. 2007: 619–47

Lunn-Rockliffe, S. (2010) 'Commemorating the usurper Magnus Maximus: Ekphrasis, poetry, and history in Pacatus' Panegyric of Theodosius', *JLA* 3: 316–36

Lynn, J. A. (2003) *Battle: A History of Combat and Culture*. Boulder

MacCormack, S. (1972) 'Change and continuity in late antiquity: the ceremony of *adventus*', *Historia* 21: 721–52

Mackay, C. (2007) *Ancient Rome: A Military and Political History*. Cambridge

MacMullen, R. (1963) *Soldier and Civilian in the Later Roman Empire*. Cambridge, MA

(1968) 'Rural Romanization', *Phoenix* 22: 337–41

(1984) 'The legion as a society', *Historia* 33: 440–56

Mann, J. C. (1983) *Legionary Recruitment and Veteran Settlement during the Principate*. London

Marichal, R. (1992) *Les Ostraca de Bu Njem*. Tripoli

Marincola, J., ed. (2007) *A Companion to Greek and Roman Historiography*. Chichester

Marshall, S. L. A. (1947) *Men against Fire*. New York

Maschek, D. (2018) 'Not *census* but *deductio*: Reconsidering the "*Ara* of Domitius Ahenobarbus"', *JRS* 108: 27–52

Mathisen, R. W. (2006) '*Peregrini*, *barbari*, and *cives Romani*: Concepts of citizenship and the legal identity of barbarians in the later Roman empire', *AHR* 111: 1011–40

(2012) 'Concepts of citizenship' in Johnson, S. F., ed., *The Oxford Handbook of Late Antiquity*, 744–63. Oxford

Mattern, S. P. (1999) *Rome and the Enemy: Imperial Strategy in the Principate*. Berkeley

Matthews, J. F. (1989) *The Roman Empire of Ammianus*. London

Mattingly, D. J. (2000) 'Whose Roman Africa?', *CR* 50: 543–5

(2011) *Imperialism, Power, and Identity: Experiencing the Roman Empire*. Princeton

(2014) 'Identities in the Roman world: Discrepancy, heterogeneity, hybridity, and plurality' in Brody & Hoffman 2014: 35–60

Maxfield, V. (1981) *The Military Decorations of the Roman Army*. London

McCall, J. B. (2002) *The Cavalry of the Roman Republic: Cavalry Combat and Elite Reputations in the Middle and Late Republic*. London

McCormick, M. (1986) *Eternal Victory: Triumphal Rulership in Late Antiquity, Byzantium and the Early Medieval West*. Cambridge

McCunn, S. (2018) 'Supply and command: A study of the late Roman commissariat.' Diss. Nottingham

McDonnell, M. (2006) *Roman Manliness:* Virtus *and the Roman Republic*. Cambridge

McEvoy, M. (2013) *Child Emperor Rule in the Late Roman West, AD 367–455*. Oxford

McLaughlin, J. J. (2014) 'Bridging the cultural divide: Libanius, Ellebichus, and letters to "barbarian" generals', *JLA* 7: 253–79

Melchior, A. (2011) 'Caesar in Vietnam: Did Roman soldiers suffer from Post-Traumatic Stress Disorder?', *G&R* 58: 209–23

Merrills, A., & Miles, R. (2010) *The Vandals*. Chichester

Messer, W. S. (1920) 'Mutiny in the Roman army: The Republic', *Classical Philology* 15: 158–75

Millar, F. (1969) 'P. Herennius Dexippus: the Greek world and the third-century invasions', *JRS* 59: 12–29

(2005) 'Last year in Jerusalem: Monuments of the Jewish War in Rome' in Edmondson, J., Mason, S., & Rives, J., eds., *Flavius Josephus and Flavian Rome*, 101–28. Oxford

Mitchell, S. (1976) 'Requisitioned transport in the Roman empire: A new inscription from Pisidia', *JRS* 66: 106–31

(1993) *Anatolia: Land, Men and Gods in Asia Minor*, 2 vols. Oxford,

Mohrmann, C. (1952) 'Encore une fois: *paganus*', *VigChr* 6: 109–21

Momigliano, A. (1975) *Alien Wisdom: The Limits of Hellenization*. Cambridge

Morony, M. (2004) 'Population transfers between Sasanian Iran and the Byzantine empire' in *La Persia e Bisanzio* (= *Atti dei Convegni Lincei* 201), 161–80. Rome

Mosci Sassi, M. G. (1983) *Il sermo castrensis*. Bologna

Müller, A. (1912) 'Das Heer Justinians', *Philologus* 71: 101–38

Nicasie, M. J. (1998) *Twilight of Empire: The Roman Army from the Reign of Diocletian until the Battle of Adrianople*. Amsterdam

Nielsen, I. (1990) Thermae et Balnea: *The Architecture and Cultural History of Roman Public Baths*. Aarhus

Nixon, C. E. V., & Saylor Rodgers, B. (1994) *In Praise of Later Roman Emperors: The* Panegyrici Latini. Berkeley

Noethlichs, K. L. (2001) 'Die "Christianisierung" des Krieges vom spätantiken bis zum frühmittelalterlichen and mittelbyzantinischen Reich', *JAC* 44: 5–22

Noreña, C. F. (2011) *Imperial Ideals in the Roman West: Representation, Circulation, Power*. Cambridge

Northwood, S. (2008) 'Census *and* tributum' in de Ligt & Northwood 2008: 257–70

Oakley, S. (1985) 'Single combat in the Roman Republic', *CQ* 35: 392–410

(1997) *A Commentary on Livy, Books VI–X. Vol. I: Introduction and Book VI*. Oxford

(1998) *A Commentary on Livy, Books VI–X. Vol. II: Books VII–VIII*. Oxford

O'Daly, G. (1999) *Augustine's* City of God: *A Reader's Guide*. Oxford

Omissi, A. (2018) *Emperors and Usurpers in the Later Roman Empire: Civil War, Panegryic and the Construction of Legitimacy*. Oxford

Onur, F. (2016) 'The Anastasian military decree from Perge in Pamphylia: Revised 2nd edition', *Gephyra* 14: 133–212

Orlin, E. M. (1997) *Temples, Religion and Politics in the Roman Republic*. Leiden

Osgood, J. (2006) *Caesar's Legacy: Civil War and the Emergence of the Roman Empire*. Cambridge

(2015) 'Ending civil war at Rome: rhetoric and reality, 88 BCE to 197 CE', *AHR* 120: 1683–95

Östenberg, I. (2009) *Staging the World: Spoils, Captives and Representations in the Roman Triumphal Procession*. Oxford

Palme, B. (2007) 'The imperial presence: government and army' in Bagnall, R. S., ed., *Egypt in the Byzantine World, 300–700*, 244–70. Cambridge

Parker, S. T. (1991) 'Preliminary report on the 1989 season of the *Limes Arabicus* Project', *BASOR, Suppl. Ser.* 27, 117–54

Paschoud, F. (2001) *Histoire Auguste*, t.V.2 (*Vies de Probus, Firmus, Saturnin, Proculus et Bonose, Carus, Numérien et Carin*). Paris

Paul, G. M. (1982) '*Urbs capta*: Sketch of an ancient literary motif', *Phoenix* 36: 144–55

(1984) *A Historical Commentary on Sallust's* Bellum Jugurthinum. Liverpool

Peachin, M. (1999) 'Five Vindolanda tablets, soldiers and the law', *Tyche* 14: 223–35

(2007) 'Petition to a centurion from the NYU papyrus collection and the question of informal adjudication performed by soldiers' in Sirks, A. J. B., & Worp, K., eds., *Papyri in Memory of P. J. Sijpesteijn*, 79–97. Chippenham

Peacock, D. P. S., & Maxfield, V.A. (1997) *Mons Claudianus: Survey and Excavation, 1987–1993*. Cairo

Pearson, E. H. (2016) 'The development of army administration in the Roman Republic.' Diss. Manchester

Petersen, L. I. R. (2013) *Siege Warfare and Military Organization in the Successor States (400–800 AD)*. Leiden

Petitjean, M. (2014) 'Classicisme, barbarie et guerre romaine: L'image du cavalier dans le monde romain tardif', *AnTard* 22: 255–62

Petrikovits, H. von (1980) 'Lixae' in Keppie & Hanson 1980: 1027–35

Pfeilschifter, R. (2007) 'The allies in the Republican army and the Romanization of Italy' in Roth, R., & Keller, R., eds., *Roman by Integration: Dimensions of Group Identity in Material Culture and Text*, 27–42. Portsmouth, RI

Phang, S. E. (2001) *The Marriage of Roman Soldiers (13 BC–AD 235): Law and Family in the Imperial Army*. Leiden

(2004) 'Intimate conquests: Roman soldiers' slave women and freedwomen', *AW* 35: 207–37

(2007) 'Military documents, languages, and literacy' in Erdkamp 2007: 286–305

(2008) *Roman Military Service: Ideologies of Discipline in the Late Republic and Early Principate*. Cambridge

Pietilä-Castrén, L. (1987) Magnificentia Publica: *The Victory Monuments of the Roman Generals in the Era of the Punic Wars*. Helsinki

Pittenger, M. R. P. (2008) *Contested Triumphs: Politics, Pageantry, and Performance in Livy's Republican Rome*. Berkeley

Pohlsander, H. (1969) 'Victory, the story of a statue', *Historia* 18: 588–97

Pollard, N. (1996) 'The Roman army as "total institution" in the Near East? Dura-Europos as a case study' in Kennedy 1996: 211–27

(2000) *Soldiers, Cities and Civilians in Roman Syria*. Ann Arbor

Popkin, M. L. (2016) *The Architecture of the Roman Triumph: Monuments, Memory and Identity*. Cambridge

Potter, D. S. (2014) *The Roman Empire at Bay, AD 180–395*, 2nd ed. London

Prag, J. R. W. (2007) '*Auxilia* and *gymnasia*: A Sicilian model of Roman Republican imperialism', *JRS* 97: 68–100

(2010) 'Troops and commanders: *auxilia externa* under the Roman Republic', *ὄρμος: Ricerche di Storia Antica* 2: 101–13

(2011) 'Provincial governors and auxiliary soldiers' in Barrandon, N., & Kirbihler, F., eds., *Les gouverneurs et les provinciaux sous la République romaine*, 15–28. Rennes

(2014) 'Bronze *rostra* from the Egadi Islands off NW Sicily', *JRA* 27: 33–59

(2015) '*Auxilia* and *clientelae*: military service and foreign *clientelae* reconsidered' in Jehne, M., & Pina Polo, F., eds., *Foreign* clientelae *in the Roman Empire: A Reconsideration*, 281–94. Stuttgart

Pringle, D. (1981) *The Defence of Byzantine Africa from Justinian to the Arab Conquest.* Oxford

Purcell, N. (1995) 'On the sacking of Carthage and Corinth' in Innes, D., Hine, H., & Pelling, C., eds., *Ethics and Rhetoric: Classical Essays for Donald Russell on his Seventy-Fifth Birthday*, 133–48. Oxford

Raaflaub, K. (1980) 'The political significance of Augustus' military reforms' in Hanson & Keppie 1980: 1005–25

(2009) 'Bellum Civile' in Griffin 2009: 175–91

Raaflaub, K., ed. (2007) *War and Peace in the Ancient World.* Malden, MA, and Oxford

Rajak, T. (2002) *Josephus: The Historian and His Society*, 2nd ed. London

Rance, P. (2000) '*Simulacra pugnae*: The literary and historical tradition of mock battles in the Roman and early Byzantine army', *GRBS* 41: 223–75

(2004a) 'The *fulcum*, the late Roman and Byzantine *testudo*: the Germanization of Roman infantry tactics?', *GRBS* 44: 265–326

(2004b) 'Drungus, *drouggos* and *drouggisti*: A Gallicism and continuity in late Roman cavalry tactics', *Phoenix* 58: 96–130

(2005) 'Narses and the Battle of Taginae (Busta Gallorum) 552: Procopius and sixth-century warfare', *Historia* 54: 424–72

(2007) 'The Later Roman Empire: Battle' in Sabin et al. 2007: vol. 2, 342–78

(2010) 'The *De Militari Scientia* or Müller Fragment as a philological resource. Latin in the East Roman army and two new loanwords in Greek: *palmarium* and *recala' *Glotta* 86: 63–92

(2014a) 'An unnoticed regimental *diaconus* in the correspondence of Theodoret of Cyrrhus', *Historia* 63: 117–28

(2014b) '*Sculca, *sculcator, exculcator* and *proculcator*: The scouts of the late Roman army and a disputed etymology', *Latomus* 73: 474–501

(2018) 'The army in peace time: The social status and function of soldiers' in Stouraitis, I., ed., *The Brill Companion to the Byzantine Culture of War, c.300–1204*, 394–439. Leiden

(2019) 'The farmer and the soldier should be friends: Justinian's legislation on the provisioning of soldiers (Novel 130)', *JLA* 12: 380–421

Rathbone, D. (2007) 'The Late Republic and the Principate: Military finance and supply' in Sabin et al. 2007: vol. 2, 158–76

(2009) 'Earnings and costs: Living standards and the Roman economy (first to third centuries AD)' in Bowman, A., & Wilson, A., eds., *Quantifying the Roman Economy: Methods and Problems*, 299–326. Oxford

Rawson, E. (1971) 'The literary sources for the pre-Marian army', *PBSR* 39: 13–31

Reddé, M. (2004) 'Réflexions critiques sur les chapelles militaires (*aedes principiorum*)', *JRA* 17: 443–62

Remijsen, S. (2007) 'The postal service and the hour as a unit of time in antiquity', *Historia* 65: 127–40

Revell, L. (2007) 'Military bath-houses in Britain: a comment', *Britannia* 38: 230–37

Ribera i Lacomba, A. (2006) 'The Roman foundation of Valencia and the town in the 2nd–1st c. BC' in Abad Casal, L., Keay, S. & Ramallo Asensio, S., eds., *Early Roman Towns in* Hispania Tarraconensis, 75–90. Portsmouth, RI

Rich, J. W. (1983) 'The supposed Roman manpower shortage of the later second century BC', *Historia* 32: 287–331

(1993) 'Fear, greed and glory: the causes of Roman war-making in the middle Republic' in Rich & Shipley 1993: 38–68

(2003) 'Augustus, war and peace' in de Blois, L. *et al.* (eds.), *The Representation and Perception of Roman Imperial Power* (= *Impact of Empire* vol. 3), 329–57. Amsterdam

(2007a) 'Warfare and the army in early Rome' in Erdkamp 2007: 7–23

(2007b) 'Tiberius Gracchus, land and manpower' in Hekster, O. *et al.*, eds., *Crises and the Roman Empire* (= *Impact of Empire* vol. 7), 155–66. Leiden

(2011) 'The *fetiales* and Roman international relations' in Richardson, J. H., & Santangelo, F., eds., *Priests and State in the Roman World*, 187–242. Stuttgart

(2012a) 'Making the emergency permanent: *auctoritas, potestas* and the evolution of the principate of Augustus' in Rivière, Y., ed., *Des réformes augustéennes*, 37–121. Rome

(2012b) 'Roman attitudes to defeat in battle under the Republic' in Simón, F. M., Pina Polo, F., & Rodríguez, J. R., eds., Vae Victis! *Perdedores en el mundo antiguo*, 83–112. Barcelona

(2013) 'Roman rituals of war' in Campbell & Tritle 2013: 542–68

(2014) 'The triumph in the Roman Republic: frequency, fluctuation and policy' in Lange, C. H., & Vervaet, F. J., eds., *The Roman Republican Triumph: Beyond the Spectacle*, 197–258. Rome

Rich, J., & Shipley, G., eds. (1993) *War and Society in the Roman World*. London

Richardson, J. H. (2018) 'P. Cornelius Scipio and the capture of New Carthage: the tide, the wind and other fantasies', *CQ* 68: 458–74

Richardson, J. S. (1986) *Hispaniae: Spain and the Development of Roman Imperialism, 218–82 BC*. Cambridge

(2012) *Augustan Rome, 44 BC to AD 14: The Restoration of the Republic and the Establishment of the Empire*. Edinburgh

Rives, J. B. (1999) 'The decree of Decius and the religion of the empire', *JRS* 89: 135–54

Robinson, H. R. (1975) *The Armour of Imperial Rome*. New York

Roller, M. B. (2001) *Constructing Autocracy: Aristocrats and Emperors in Julio-Claudian Rome*. Princeton

Roselaar, S. T. (2010) *Public Land in the Roman Republic: A Social and Economic History of the* ager publicus *in the Roman Republic, 396–89 BC*. Oxford

Rosenstein, N. (1990) Imperatores Victi: *Military Defeat and Aristocratic Competition in the Middle and Late Republic*. Berkeley

(2004) *Rome at War: Farms, Families, and Death in the Middle Republic*. Chapel Hill, NC

(2007) 'War and peace, fear and reconciliation at Rome' in Raaflaub 2007: 226–44
(2012a) *Rome and the Mediterranean 290 to 146 BC: The Imperial Republic.* Edinburgh
(2012b) 'Integration and armies in the Middle Republic' in Roselaar, S. T., ed., *Processes of Integration and Identity Formation in the Roman Republic*, 85–103. Leiden
Ross, A. J. (2016) *Ammianus' Julian: Narrative and Genre in the* Res Gestae. Oxford
Roth, J. P. (1994) 'The size and organization of the Roman imperial legion', *Historia* 43: 346–62
(1999) *The Logistics of the Roman Army at War (264 BC–AD 235).* Leiden
(2009) *Roman Warfare.* Cambridge
(2016) 'Josephus as a military historian' in Chapman, H. H., & Rodgers, Z., eds., *A Companion to Josephus*, ch. 11. Chichester
Roueché, C. (1981) 'Rome, Asia and Aphrodisias in the third century', *JRS* 71: 103–20
Rüger, C. B. (1987), 'Beobachtungen zu den epigraphischen Belegen der Muttergottheiten in den lateinischen Provinzen des Imperium Romanum', in Bauchhenß , G., & Neumann, G., eds., *Matronen und verwandte Gottheiten*, 1–30. Cologne
Rüpke, J. (1990) *Domi militiae: Die religiöse Konstruktion des Krieges in Rom.* Stuttgart
Russell, D. A., & Wilson, N. G. (1981) *Menander Rhetor: A Commentary.* Oxford
Sabin, P. (1996) 'The mechanics of battle in the Second Punic War' in Cornell et al. 1996: 59–80
(2000) 'The face of Roman battle', *JRS* 90: 1–17
(2007) 'The Hellenistic world and Roman Republic: Battle – A. Land battles' in Sabin et al. 2007: vol. 1, 399–433
Sabin, P., van Wees, H., & Whitby, M., eds. (2007) *The Cambridge History of Greek and Roman Warfare*, 2 vols. Cambridge
Saddington, D. B. (1982) *The Development of the Roman Auxiliary Forces from Caesar to Vespasian, 49 BC–AD 79.* Harare
Saller, R. P. (1994) *Patriarchy, Property and Death in the Roman Family.* Cambridge
(2002) 'Framing the debate over growth in the ancient economy' in Scheidel & von Reden 2002: 251–69
Salway, B. (2014) 'Late antiquity' in Bruun & Edmondson 2014: 364–93
Salzman, M. R. (2006) 'Symmachus and the "barbarian" generals', *Historia* 55: 352–67
Scheidel, W. (1996) 'The demography of the Roman imperial army' in his *Measuring Sex, Age and Death in the Roman Empire: Explorations in Ancient Demography*, 93–138. Ann Arbor
Scheidel, W., ed. (2001a) *Debating Roman Demography.* Leiden

(2001b) 'Progress and problems in Roman demography' in Scheidel 2001a: 1–81
(2007) 'Demography' in Scheidel, Morris & Saller 2007: 38–86
(2008) 'Roman population size: the logic of the debate' in de Ligt & Northwood 2008: 17–70
Scheidel, W., ed. (2018) *The Science of Roman History: Biology, Climate and the Future of the Past*. Princeton
Scheidel, W., Morris, I. & Saller, R., eds. (2007) *The Cambridge Economic History of the Greco-Roman World*. Cambridge
Scheidel, W., & von Reden, S., eds. (2002), *The Ancient Economy*. Edinburgh
Sekunda, N. (2001) *Hellenistic Infantry Reform in the 160s BC*. Lodz
(2007) 'The Hellenistic world and Roman Republic: Land forces' in Sabin, Van Wees & Whitby 2007: vol. 1, 325–57
Sessa, K. (2019) 'The new environmental fall of Rome: a methodological consideration', *JLA* 12: 211–55
Seston, W. (1969) 'Feldzeichen', *RAC* 7: 689–711
Shackelton Bailey, D. R. (1977) *Cicero: Epistulae ad Familiares*, 2 vols. Cambridge
Shaw, B. D. (1983) 'Soldiers and society: the army in Numidia', *OPVS* 2: 133–59
(1999) 'War and violence' in G. W. Bowersock, P. Brown & O. Grabar, eds., *Late Antiquity: A Guide to the Postclassical World*, 130–69. Cambridge, MA
Sidebottom, H. (2004) *Ancient Warfare: A Very Short Introduction*. Oxford
(2005) 'Roman imperialism: the changed outward trajectory of the Roman empire', *Historia* 54: 315–30
Sidebottom, H., & Whitby, M., eds. (2018) *The Wiley-Blackwell Encyclopaedia of Ancient Battles*. Chichester
Smith, C. J. (1998) 'Onasander on how to be a general' in Austin, M., Harries, J., & Smith, C., eds., Modus Operandi: *Essays in Honour of Geoffrey Rickman*, 151–66. London
Southern, P. (1989) 'The *numeri* of the Roman imperial army', *Britannia* 20: 81–140
Speidel, M. A. (2001) 'Specialisation and promotion in the Roman imperial army' in L. de Blois, ed., *Administration, Prosopography and Appointment Policies in the Roman Empire*. 50–61. Amsterdam
(2009) 'Les longues marches des armées romaines. Reflets épigraphiques de la circulation des militaires dans la province d'Asie au IIIe siècle apr. J.C.', *Cahiers du Centre Gustav Glotz* 20: 199–210
(2012a) 'Being a soldier in the Roman imperial army – expectations and responses' in C. Wolff, ed., *Le métier de soldat dans le monde romain*, 175–93. Lyons
(2012b) 'Dressed for the occasion: Clothes and context in the Roman army' in M.-L. Nosch, ed., *Wearing the Cloak: Dressing the Soldier in Roman Times*, 1–12. Oxford
(2014) 'The Roman army' in Bruun & Edmondson 2014: 319–44
(2018) 'Soldiers and documents: insights from Nubia. The significance of written documents in Roman soldiers' everyday lives' in Kolb, A., ed., *Literacy in Ancient Everyday Life*, 179–200. Berlin

Speidel, M. P. (1973) 'The pay of the *auxilia*', *JRS* 63: 141–7

(1978) *The Religion of Iuppiter Dolichenus in the Roman Army*. Leiden

(1984) '*Catafractarii, clibanarii* and the rise of the later Roman mailed cavalry', *Epigraphica Anatolica* 3–4: 151–6

(1985) 'Furlough in the Roman army', *YCS* 28: 283–93

(1989) 'The soldiers' servants', *Ancient Society* 20: 239–48

Speidel, M. P., & Dimitrova-Milčeva, A. (1978) 'The cult of the *genii* in the Roman army and a new military deity', *ANRW* II.16.2, 1542–55

Spiller, R. J. (1988) 'S. L. A. Marshall and the ratio of fire', *R.U.S.I. Journal* 133: 63–71

Stauner, K. (2004) *Das offizielle Schriftwesen des römischen Heeres von Augustus bis Gallienus (27 v.Chr.–268 n.Chr.): Eine Untersuchung zu Struktur, Funktion und Bedeutung der offiziellen militärischen Verwaltungsdokumentation und zu deren Schreibern*. Bonn

Steel, C. (2013) *The End of the Roman Republic, 146 to 44 BC: Conquest and Crisis*. Edinburgh

Stickler, T. (2007) 'The foederati' in Erdkamp 2007: 495–514

Stoll, O. (2001) *Zwischen Integration und Abgrenzung: Die Religion des römischen Heeres im Nahen Osten*. St Katharinen

(2007) 'The religions of the armies' in Erdkamp 2007: 451–76

(2009) 'Integration und doppelte Identität: Römisches Militär und die Kulte der Soldaten und Veteranen in Ägypten von Augustus bis Diokletian', in Gundlach, R., & Vogel, C., eds., *Militärgeschichte des pharaonischen Ägypten*, 419–58. Paderborn

Strachan, H. (2006) 'Training, morale and modern war', *Journal of Contemporary History* 41: 211–27

Swain, S., & Edwards, M., eds. (2004) *Approaching Late Antiquity: The Transformation from Early to Late Empire*. Oxford

Swift, L. J. (2007) 'Early Christian views on violence, war, and peace' in Raaflaub 2007: 279–96

Tan, J. (2017) *Power and Public Finance at Rome, 264–49 BCE*. Oxford

Thorburn, J. E. (2003) '*Lixae* and *calones*: following the Roman army', *Classical Bulletin* 79: 47–61

Todd, M. (1996) 'Review of Künzl 1993', *Britannia* 27: 481–2

Tomlin, R. S. O. (1998) 'Christianity and the Roman army' in Lieu, S. N. C., & Montserrat, D., eds., *Constantine: History, Hagiography and Legend*, 21–51. London

Töpfer, K. M. (2011) *Signa Militaria: Die römischen Feldzeichen in der Republik und im Prinzipat*. Mainz

Treadgold, W. (1995) *Byzantium and its Army, 284–1081*. Stanford

(2007) *The Early Byzantine Historians*. London

Trombley, F. R., & Watt, J. W. (2000) *The Chronicle of Pseudo-Joshua the Stylite*. Liverpool

Tusa, S., & Royal, J. (2012) 'The landscape of the naval battle at the Egadi Islands (241 BC)', *JRA* 25: 7–48

Vanhoutte, S. (2015) 'The late Roman coastal fort of Oudenburg (Belgium): Spatial and functional transformations within the fort walls' in Collins, R., Symonds, M., & Weber, M., eds., *Roman Military Architecture on the Frontiers: Armies and their Architecture in Late Antiquity*, 62–75. Oxford
Van Nuffelen, P. (2012) *Orosius and the Rhetoric of History*. Oxford
Van Wees, H. (1997) 'The Seleucid army', *CR* 47: 356–7
Walters, J. (1997) 'Invading the Roman body: Manliness and impenetrability in Roman thought' in Hallett, J. P., & Skinner, M. B., eds., *Roman Sexualities*, 29–43. Princeton
Walzer, M. (2015) *Just and Unjust Wars: A Moral Argument with Historical Illustrations*, 5th ed. New York
Ward, G. (2016) 'The Roman battlefield: Individual exploits in warfare of the Roman Republic' in Reiss, W., & Fagan, G. G., eds., *The Topography of Violence in the Greco-Roman World*, 299–324. Ann Arbor
Watt, J. W. (1999) 'Greek historiography and the *Chronicle* of Joshua the Stylite' in Reinink, G.J., & Klugkist, A.C., eds., *After Bardaisan: Studies on Continuity and Change in Syriac Christianity in Honour of Prof. Han J.W. Drijvers*. 317–28. Louvain
Weinstock, S. (1957) 'Victor and invictus', *HThR* 50: 211–47
(1960) 'Pax and the *Ara Pacis*', *JRS* 50: 44–58
Weiss, P. (2003) 'The vision of Constantine', *JRA* 16: 237–59
Welles, C. B., Fink, R. O., & Gilliam, J. F., eds. (1959) *The Excavations at Dura-Europos, Final Report V, Part I (The Parchments and Papyri)*. New Haven
Wells, C. M. (1989) 'Celibate soldiers: Augustus and the army', *AJAH* 14: 180–90
Wells, P. S. (2003) *The Battle That Stopped Rome: Emperor Augustus, Arminius, and the Slaughter of the Legions in the Teutoburg Forest*. New York
Welwei, K.-W. (1988) *Unfreie im antiken Kriegsdienst*, Teil 3: *Rom*. Stuttgart
Whately, C. (2016) *Battles and Generals: Combat, Culture and Didacticism in Procopius'* Wars. Leiden
(2018) 'Combat motivation and cohesion in the age of Justinian' in Greatrex, G., & Janniard, S., eds., *Le monde de Procope/The World of Procopius*, 185–203. Paris
(forthcoming) 'Women and the military in the age of Justinian'
Wheeler, E. L. (1979) 'The legion as phalanx', *Chiron* 9: 303–18
(1988) 'Πολλὰ κενὰ τοῦ πολέμου: The history of a Greek proverb', *GRBS* 29: 153–84
(1996) 'The laxity of the Syrian legions' in Kennedy 1996: 229–76
(2001) 'Firepower: Missile weapons and the "face of battle"', *Electrum* 5: 169–84
(2004) 'The legion as phalanx in the Late Empire (Part 1)' in Le Bohec, Y., & Wolff, C., eds., *L'Armée romaine de Dioclétien à Valentinien Ier*, 309–58. Lyon
(2008) '*Pullarii, Marsi, haruspices*, and *sacerdotes* in the Roman imperial army' in Schellenburg, H. M., Hirschmann, V. E., & Krieckhous, A., eds.,

A Roman Miscellany: Essays in Honour of Anthony R. Birley on His Seventieth Birthday, 185–201. Gdansk

(2009) 'Shock and awe: Battles of the gods in Roman imperial warfare' in Wolff 2009: 225–67

(2010) 'Polyaenus: *scriptor militaris*' in Brodersen, K., ed., *Polyainos: Neue Studien*, 7–54. Berlin

Whitby, M. (1988) *The Emperor Maurice and His Historian: Theophylact Simocatta on Persian and Balkan Warfare.* Oxford

(1995) 'Recruitment in Roman armies from Justinian to Heraclius (*ca.* 565–615)' in Cameron, Averil, ed., *The Byzantine and Early Islamic Near East, III: States, Resources and Armies*, 61–124. Princeton

(2004) 'Emperors and armies, AD 235–395' in Swain & Edwards 2004: 156–86

(2007) 'Reconstructing ancient warfare' in Sabin et al. 2007: vol. 1, 54–81

Whitmore, A. M. (2013) 'Small finds and the social environment of the Roman baths.' Diss. Iowa (https://ir.uiowa.edu/etd/1797)

Wickham, C. (2005) *Framing the Early Middle Ages: Europe and the Mediterranean, 400–800.* Oxford

Wienand, J. (2015) '*O tandem felix civili, Roma, Victoria!* Civil-war triumphs from Honorius to Constantine and back' in Wienand, J., ed., *Contested Monarchy: Integrating the Roman Empire in the Fourth Century AD*, 169–97. Oxford

Wilkinson, K. W. (2012) 'Aurelius Gaius (*AE* 1981.777) and imperial journeys 293–299,' *ZPE* 183: 53–58

Willcock, M. M. (1995) *Cicero: The Letters of January to April 43 BC.* Warminster

Williams, C. (2008) 'Review of McDonnell 2006', *JRS* 98: 204–5

Wolff, C., ed. (2009) *L'Armée romaine et la religion sous le haut-empire romain.* Lyons

Woodman, A. J. (1998) 'Introduction: The literature of war' in his *Tacitus Reviewed*, 1–20. Oxford

Woolf, G. (1993) 'Roman peace' in Rich & Shipley 1993: 171–94

(1998) *Becoming Roman: The Origins of Provincial Civilization in Gaul.* Cambridge

(2012) *Rome: An Empire's Story*. Oxford

Wright, D. H. (1986) 'Justinian and an archangel' in Feld, O., & Peschlow, U., eds., *Studien zur spätantiken und byzantinischen Kunst: Friedrich Wilhelm Deichmann gewidmet*, 75–9. Bonn

Zhmodikov, A. (2000) 'Roman Republican heavy infantrymen in battle (fourth–second centuries BC)', *Historia* 49: 67–78

Zienkiewicz, J. D. (1986) *The Legionary Fortress Baths at Caerleon.* Cardiff

Ziolkowski, A. (1993) '*Urbs direpta*, or how the Romans sacked cities' in Rich & Shipley 1993: 69–91

Index

Page numbers in *italics* indicate figures.

Lightning Source UK Ltd.
Milton Keynes UK
UKHW020018140920
369819UK00007B/56

9 781107 638280